Unconventional Techniq[ues...]
Light Alloys and Compo[sites...]

Jose Martin Herrera Ramirez
Raul Perez Bustamante
Cesar Augusto Isaza Merino
Ana Maria Arizmendi Morquecho

Unconventional Techniques for the Production of Light Alloys and Composites

Jose Martin Herrera Ramirez
Advanced Materials Research Center
Chihuahua, Mexico

Cesar Augusto Isaza Merino
Faculty of Engineering, GIIEN Research Group
Pascual Bravo University Institution
Medellin, Colombia

Raul Perez Bustamante
Mexican Corporation for Research on Materials
CONACYT
San Luis Potosi, Mexico

Ana Maria Arizmendi Morquecho
Advanced Materials Research Center
Monterrey, Mexico

ISBN 978-3-030-48124-7   ISBN 978-3-030-48122-3   (eBook)
https://doi.org/10.1007/978-3-030-48122-3

© Springer Nature Switzerland AG 2020

This work is subject to copyright. All rights are reserved by the Publisher, whether the whole or part of the material is concerned, specifically the rights of translation, reprinting, reuse of illustrations, recitation, broadcasting, reproduction on microfilms or in any other physical way, and transmission or information storage and retrieval, electronic adaptation, computer software, or by similar or dissimilar methodology now known or hereafter developed.

The use of general descriptive names, registered names, trademarks, service marks, etc. in this publication does not imply, even in the absence of a specific statement, that such names are exempt from the relevant protective laws and regulations and therefore free for general use.

The publisher, the authors, and the editors are safe to assume that the advice and information in this book are believed to be true and accurate at the date of publication. Neither the publisher nor the authors or the editors give a warranty, expressed or implied, with respect to the material contained herein or for any errors or omissions that may have been made. The publisher remains neutral with regard to jurisdictional claims in published maps and institutional affiliations.

This Springer imprint is published by the registered company Springer Nature Switzerland AG
The registered company address is: Gewerbestrasse 11, 6330 Cham, Switzerland

To my wife Ana Maria and my son Martin Alejandro... thanks for all your love, support, and patience toward me. Once again I was absent from you, for which I apologize. I also thank my maternal family, my mother, my siblings, nieces, and nephews, for the support they provided me, in one way or another, throughout my personal and professional life.
Dr. Jose Martin Herrera Ramirez

This book is dedicated to my parents and brother with love and gratitude.
Dr. Raul Perez Bustamante

This book is dedicated to my wife, who helped me through this process; her constant encouragement provided me the strength to complete this task. It is also dedicated to my daughter Abril, who is the new motivation for my life.
Dr. Cesar Augusto Isaza Merino

This book is dedicated with love to my father, who told me the stories that matter, and to my mother, who taught me that to achieve success one has to work hard and with determination to get the goals, regardless the result, always give my best.
Dr. Ana Maria Arizmendi Morquecho

Preface

The transport industry in its various modalities constantly requires the development of advanced materials and new technologies for their application. The common objective is energy saving. For this purpose, light alloys and composites emerge as a solution that is constantly adjusted in compliance with such requirements. This is carried out through a determined research, where the limits of the physical properties of light alloys and composites are explored through their chemical alteration, the addition of micro- and nano-agents or reinforcements, the application of heat treatments, the use of unconventional and cutting-edge technologies, and even a combination of all the previous routes.

Books and publications on the importance of the material progress in optimizing their properties are available in the literature. However, the present book provides the latest advances on the manipulation of the physical properties of light alloys and composites, in which unconventional methods and vanguard technologies are contemplated for their enhanced performance.

This book presents an overview of light alloys and composites, as well as their manufacturing process, considering non-traditional techniques in their production as forefront technologies. Their respective characterization techniques and applications in the aeronautical and aerospace industries will be covered in the following chapters:

- Introduction
- Manufacturing Processes of Light Metals and Composites
- Powder Metallurgy
- Sandwich Technique
- Severe Plastic Deformation
- Additive Manufacturing
- Thermal Spray Coatings
- Characterization Techniques
- Interface Characterization of Light Metal Matrix Composites
- Applications of Light Alloys and Composites in the Aeronautical and Aerospace Industries

Several case studies are presented in this book, where aluminum, magnesium, titanium, and their alloys receive special attention due to their high interest in the transportation industry. The advanced materials rely on the use of various techniques and reinforcing agents to increase the physical properties of the resulting composite materials.

Advanced Materials Research Center Chihuahua, Mexico	Jose Martin Herrera Ramirez
Mexican Corporation for Research on Materials CONACYT San Luis Potosi, Mexico	Raul Perez Bustamante
Faculty of Engineering, GIIEN Research Group Pascual Bravo University Institution Medellin, Colombia	Cesar Augusto Isaza Merino
Advanced Materials Research Center Monterrey, Mexico	Ana Maria Arizmendi Morquecho

Acknowledgments

This book refers to works made in several research establishments but could not have been written without the contributions of a considerable number of research students and technicians who have worked with the authors. Some of these people are mentioned in the references but by no means all. Thanks must go to all of them for throwing light on these fascinating and difficult subjects. Special thanks to MSc Jose Ernesto Ledezma-Sillas for designing most of the 3D model images and for having helped in the book's revision.

Dr. Martin Herrera and Dr. Ana Arizmendi thank Advanced Materials Research Center (CIMAV) and National Council of Science and Technology (CONACYT). They are grateful to the following people for the technical support provided during the synthesis and characterization of samples: Dr. Nayeli Pineda-Aguilar, Dr. Cesar Cutberto Leyva-Porras, Dr. Caleb Carreño-Gallardo, MSc Miguel Esneider-Alcala, MSc Raul Armando Ochoa-Gamboa, MSc Karla Campos-Venegas, MSc Erika Ivonne Lopez Martinez, MSc Roberto Pablo Talamantes-Soto, MSc Carlos Elias Ornelas-Gutierrez, MSc Oscar Omar Solis-Canto, MSc Daniel Lardizabal-Gutierrez, and MSc Armando Tejeda Ochoa.

Dr. Raul Perez thanks The Mexican Corporation for Research on Materials (COMIMSA), The Advanced Technology Center (CIATEQ), Cummins Inc., Renishaw Mexico, S. de R.L. de C.V.; and the project 850 of the program Catedras CONACYT.

Dr. Cesar Isaza would like to extend sincere gratitude to Dr. Juan Manuel Meza Meza and to National University of Colombia for allowing him the use of their facilities. Thanks are also due to Dr. Jose Miguel Yacaman for allowing him the use of the Kleberg Microscopy Laboratory Center facilities, as well as all researchers from the Department of Physics and Astronomy for helping him during his stay at the University of Texas-San Antonio. Finally, he thanks Pascual Bravo University Institute for the financial support through the Institutional Call for Research Projects and/or Artistic Creation under the resolution 067 of January 30, 2019.

Dr. Ana Arizmendi thanks Service and Specialties and Eutectic Mexico for having allowed access and use of their facilities and providing some images used in this book.

Authorship Statement

The authors claim the following authorship in this book:

Dr. Jose Martin Herrera Ramirez, Dr. Raul Perez Bustamante, Dr. Cesar Augusto Isaza Merino, and Dr. Ana Maria Arizmendi Morquecho are contributing authors. All authors proofread the entire book.

Contents

1	**Introduction**		1
	1.1 Aluminum		2
	1.2 Magnesium		3
	1.3 Titanium		3
	1.4 Beryllium		4
	1.5 Metal Matrix Composites		4
	1.6 Reinforcing Materials		4
	1.7 Production Techniques of Light Alloys and Composites		5
	1.8 Powder Metallurgy		6
	1.9 Mechanical Alloying and Milling		6
	1.10 Sandwich Technique		6
	1.11 Severe Plastic Deformation		7
	1.12 Additive Manufacturing		7
	1.13 Thermal Spray Coatings		8
	1.14 Characterization Techniques		8
	1.15 Mechanical Properties		9
	1.16 Applications of Light Alloys and Composites		9
	References		10
2	**Manufacturing Processes of Light Metals and Composites**		13
	2.1 Introduction to Manufacturing Process		13
	2.2 Unconventional Manufacturing Processes for Nanocomposites		19
		2.2.1 Stir Casting	20
		2.2.2 Semisolid Metal	20
		2.2.3 Selective Laser Melting	20
	2.3 Design of New Alloys		21
	2.4 Conclusions		30
	References		30
3	**Powder Metallurgy**		33
	3.1 Introduction		33
	3.2 Advantages of Powder Metallurgy		35

	3.3	Metal Powder Materials for the Development of Light Alloys	35
	3.4	Consolidation of Metal Powders	37
	3.5	Sintering of Metal Powders	38
	3.6	Powder Metallurgy and Mechanical Alloying in the Production of Light Alloys Strengthened with Carbon Nanotubes	40
	3.7	Conclusions	46
	References		46
4	**Sandwich Technique**		49
	4.1	Introduction to the Sandwich Technique	49
	4.2	Magnesium Alloys	50
	4.3	Aluminum Alloys	51
	4.4	Metal Matrix Composites Synthesis	51
		4.4.1 Polymer Matrix Composites Synthesis	51
		4.4.2 Metal Matrix Composite Synthesis	53
		4.4.3 Diffusion Bonding Mechanism	54
		4.4.4 Microstructural and Structural Analysis of Composites	58
		4.4.5 Dispersion Quantification of MWCNTs in Metal Matrix Composites Fabricated by the Sandwich Technique	61
		4.4.6 Microstructural Evolution Between Metallic Sheets in the Composites	62
		4.4.7 Bulk Mechanical Properties of Metal Matrix/MWCNTs Composites	63
	4.5	Conclusions	65
	References		66
5	**Severe Plastic Deformation**		69
	5.1	Introduction to Severe Plastic Deformation	69
	5.2	High-Pressure Torsion	69
	5.3	Titanium-Magnesium Alloys	71
	5.4	Ti-Mg Alloys Synthesized by HPT	72
	5.5	Characterization of Ti-Mg Alloys	76
	5.6	Conclusions	86
	References		86
6	**Additive Manufacturing**		89
	6.1	Introduction to Additive Metal Manufacturing	89
	6.2	Characteristics and Advantages of Additive Metal Manufacturing	91
	6.3	Technologies Used in the AM of Light Alloys and Composites	94
		6.3.1 Direct Metal Laser Sintering	94
		6.3.2 Process and Equipment Used in DMLS	95
		6.3.3 Powders Used in DMLS	95
		6.3.4 Use of DMLS in the Plastic Injection Mold Industry	97

		6.4	Cold Spray Low Pressure................................	98
			6.4.1 Process and Equipment Used in CSLP	98
			6.4.2 Powders Used in CSLP	99
		6.5	Conclusions	101
		References...		101
7	**Thermal Spray Coatings**			103
	7.1	Introduction to Thermal Spray Processes		103
	7.2	Thermal Spray Processes Used to Coat Light Alloys		106
		7.2.1 Combustion Flame Spray...........................		107
		7.2.2 Arc Spray.......................................		108
		7.2.3 High Velocity Oxy-Fuel..........................		109
		7.2.4 Cold and Warm Spray		110
	7.3	Spray Materials for Modification of Light Alloys..............		112
	7.4	Microstructure of Coatings		115
	7.5	Conclusions		123
	References...			125
8	**Characterization Techniques**...............................			129
	8.1	Introduction to Characterization Techniques.................		129
	8.2	Chemical Analysis.......................................		130
	8.3	Thermal Analysis.......................................		131
	8.4	Density Measurement		131
		8.4.1 Archimedes' Method		132
		8.4.2 Pycnometry		133
		8.4.3 Computed Tomography Scanning		135
	8.5	Optical Microscopy......................................		136
	8.6	X-Ray Diffraction		136
	8.7	Raman Spectroscopy		138
	8.8	Scanning Electron Microscopy and Energy-Dispersive Spectroscopy		140
	8.9	TEM Sample Preparation.................................		143
		8.9.1 Electropolishing.................................		143
		8.9.2 Ultramicrotomy		144
		8.9.3 Focused Ion Beam		144
	8.10	Transmission Electron Microscopy		145
	8.11	High-Resolution Transmission Electron Microscopy		147
	8.12	Electron Energy Loss Spectroscopy (EELS).................		149
	8.13	X-Ray Photoelectron Spectroscopy		150
	8.14	Mechanical Properties		151
		8.14.1 Bulk-Scale Mechanical Testing....................		152
		8.14.2 Nano- and Micromechanical Testing................		155
	8.15	Conclusions		162
	References...			163

9	**Interface Characterization**...		167
	9.1	Introduction to the Interface Behavior Between Metal Matrix and Reinforcement.....................................	167
	9.2	Interface Strengthening Mechanisms in Light Metallic Materials	169
	9.3	Other Interactions Between CNTs and Metal Matrix	170
	9.4	Interface Characterization in Metal Matrix Composites	171
		9.4.1 Interface Characterization in Composites Fabricated by Sandwich Technique...........................	171
		9.4.2 TEM Analysis of Metal Matrix Composites	173
		9.4.3 HRTEM Analysis at the Interface Between MWCNTs and Metal Matrix	174
		9.4.4 Elemental Analysis and Energy Loss Spectroscopy at the Interface Between MWCNTS and Metal Matrix....	179
	9.5	Conclusions	180
	References...		181
10	**Applications in the Aeronautical and Aerospace Industries**		183
	10.1	Introduction to Industrial Applications of Light Alloys and Composites	183
	10.2	Applications of Thermal Spray Coatings....................	185
	10.3	Conclusions	194
	References...		195
Index			197

About the Authors

Dr. Martin Herrera completed his BSc in Chemical Engineering at Military School of Engineers (University of the Mexican Army and Air Force). Then he studied a MSc in Metallurgical Engineering at National Polytechnic Institute. Later, he earned his PhD in Materials Science and Engineering from National School of Mines of Paris. After graduation, Dr. Herrera worked for the Army at the General Directorate of Military Industry, where he held various positions including the head and project leader of the Applied Research Center and Technology Development for the Military Industry (CIADTIM). Retiring from the Army, Dr. Herrera joined Advanced Materials Research Center (CIMAV) in Chihuahua, Mexico, as a full-time researcher. His current research focuses mainly on the development of metallic alloys and composites. Dr. Herrera has authored or co-authored around 140 publications including journal papers, proceedings, books and chapters, as well as basic science and technological projects. He has been thesis advisor to doctoral, master's, and bachelor's students.

Dr. Raul Perez-Bustamante has a BSc degree in Mechanical Engineering from the Technological Institute of Chihuahua (ITCH), Mexico. He completed his MSc degree and Doctoral degree in Material Science at Advanced Materials Research Center (CIMAV), Mexico. Since completing his doctorate, Dr. Perez-Bustamante has held the position of Research Professor at The Mexican Corporation for Research on Materials (COMIMSA), as part of the program 850 of CATEDRAS of The National Council for Science and Technology (CONACYT). He is an enthusiastic researcher in the study of the mechanical and microstructural behavior of metal matrix composite materials modified by dispersing micro- and nanoparticles. Specifically, light alloys present an attractive study opportunity due to their versatility in terms of mechanical strength and lightness. The dispersion of different nature nanoparticles in this type of alloys provides a window to observe their behavior and effect on different light alloys, particularly when alternative processes are used in their synthesis.

Dr. Cesar Isaza completed his bachelor's degree in Mechanical Engineering and his master's degree in Engineering at the National University of Colombia-Medellin (UNAL). His research work focused on the study of metal matrix composites; this work was enough for obtaining meritorious mention by such university. His doctorate continued at UNAL supported by the National Doctoral Scholarship granted by the Republic of Colombia, earning the highest honorable mention (summa cum laude) granted by UNAL for the contribution to the development of metal matrix composites. After completing his doctorate, Dr. Isaza held the position of Research Professor at the Pascual Bravo University Institution-Medellin. Since then, he has participated in different research projects with UNAL, Pascual Bravo University Institution, and the Center for Research in Advanced Materials Research Center (CIMAV). His research results have been disclosed through more than 50 publications in journals, proceedings, and research reports, among others. Additionally, Dr. Isaza is advising postgraduate students in the area of hybrid metal matrix composites.

Dr. Ana Arizmendi has a BSc degree in Materials Engineering from Technological Institute of Saltillo, Mexico. She completed her MSc degree in Metallurgical Engineering Sciences and Doctoral degree in Sciences in Metallurgical and Ceramic Engineering at CINVESTAV-IPN, Mexico. After graduating, her main experience in the industry was as product innovation engineer at an automotive company. After that, she worked for 2 years as a researcher at Mexican Corporation for Research on Materials (COMIMSA), a Center belonging to the National Council for Science and Technology (CONACYT). Since 2008 she has been working at Advanced Materials Research Center (CIMAV) in Monterrey as a full-time researcher and recently she completed a MSc in Commercialization of Science and Technology (CIMAV-University of Texas, USA, joint program). Her main research line at CIMAV is the design, development and characterization of metal matrix nanocomposites and nanostructured coatings with emphasis on industrial applications. Her current activities include the development of frontier and applied research projects, authorship of scientific papers, and the formation of specialized human resources. She has actively participated in the Innovation Binational Node Program of the Mexico North Region, which is affiliated to iCorps and the NSF-USA. She is an active member of the Nanotechnology Cluster of Nuevo Leon State and participates in the International Electrophoretic Deposition Network.

Abbreviations

AAS	Atomic absorption spectrometry
ADF	Annular dark-field
AFM	Atomic force microscopy
AFNOR	Association Française de Normalisation
AISI	American Iron and Steel Institute
AM	Additive manufacturing
ANSI	American National Standards Institute
ARB	Accumulative roll-bonding
AS	Standards Australia
AS	Arc spray
ASME	American Society of Mechanical Engineers
ASTM	American Society for Testing and Materials
BS	British Standards
BSE	Backscattered electron
CALPHAD	CALculation of PHAse Diagram
CCDF	Cyclic closed-die forging
CEC	Cyclic extrusion and compression
CFD	Computational fluid dynamics
CFS	Combustion flame spray
CNT	Carbon nanotube
CRSSs	Critical resolved shear stresses
CS	Cold spray
CSLP	Cold spray low pressure
CT	Computed tomography
CVD	Chemical vapor deposition
DIN	Deutsches Institut für Normung
DMLS	Direct metal laser sintering
DSC	Differential scanning calorimetry
DTA	Differential thermal analysis
ECAP	Equal-channel angular pressing
EDS	Energy dispersion spectroscopy
EELS	Electron energy loss spectroscopy

FAAS	Flame atomic absorption spectrometry
f-CNT	Functionalized carbon nanotube
FE-SEM	Field emission scanning electron microscopy
FFT	Fast Fourier transform
f-G	Functionalized graphene
FIB	Focused ion beam
HFIS	High frequency plasma or induction systems
HPT	High-pressure torsion
HRTEM	High-resolution transmission electron microscopy
HV	Vickers microhardness
HVOF	High velocity oxy-fuel
ICDD	International Centre for Diffraction Data
ICP	Inductively coupled plasma
ICP-MS	Inductively coupled plasma mass spectrometry
ICP-OES	Inductively coupled plasma optical emission spectrometry
JIS	Japanese Industrial Standards
MA	Mechanical alloying
MDF	Multi-directional forging
MM	Mechanical milling
MMCs	Metal matrix composites
MP-HPS	Multi-pass high-pressure sliding
MWCNTs	Multiwalled carbon nanotubes
OM	Optical microscopy
PCA	Process control agent
PEO	Plasma electrolytic oxide
PM	Powder metallurgy
P-T	Pressure-temperature
PVA	Polyvinyl alcohol
RCS	Repetitive corrugation and straightening
SAED	Selected area electron diffraction
SEM	Scanning electron microscopy
SE-SEM	Secondary electron scanning electron microscopy
SFM	Scanning force microscopy
SLM	Selective laser melting
SPD	Severe plastic deformation
SSM	Semi-solid metal
SSMR	Super short multi-pass rolling
STEM	Scanning transmission electron microscopy
STS	Severe torsion straining
SWCNTs	Single-walled carbon nanotubes
TE	Twist extrusion
TEM	Transmission electron microscopy
TGA	Thermogravimetric analysis
XPS	X-rsay photoelectron spectroscopy
XRD	X-ray diffraction

Chapter 1
Introduction

Abstract The constant innovation of the modern aeronautical and aerospace industries demands the use of better and lighter materials, which represents the most efficient way to reduce the weight of structural components and devices. To achieve this, increasing the resistance-weight ratio implies the use of improved techniques and processing methods for the component manufacturing, which are mainly mass-produced from light alloys and composites, directly impacting the best aircraft performance. This chapter is dedicated to provide a brief description of various types of lightweight materials and composites currently in use, which have been shown to be able of conferring improved properties when they are produced by unconventional processing techniques. For composites materials, the chapter describes some of the most used reinforcement constituents for industrial applications. A brief explanation of various processes for manufacturing lightweight materials and composites, as well as some conventional and sophisticated characterization techniques to evaluate them is afforded.

The high-performance characteristics of a modern aeronautic industry are a direct consequence of the high-performance light materials, composites, and their manufacturing. For example, a commercial aircraft will fly over 60,000 h during its life years [1]; this time life suggests to use metallic materials and composites with high mechanical properties, fracture toughness, fatigue resistance, corrosion resistance, etc. Therefore, designers around the world are constantly searching for lighter and stronger materials. The use of lighter materials is the most efficient way of reducing weight in structural devices of aircrafts, i.e., increasing the resistance-weight ratio (specific properties) [2].

Since the 1920s, aluminum has been the metallic material more used in airframes and in some other structural devices. However, between the years 1960 and 1970, other high-performance materials such as composites started to be used in some military industry applications (e.g., F14 and F15 military aircrafts). Boron- and carbon-epoxy composites were the first composites used in the aircraft military industry, whose application resulted in significant weight savings (20%). Nowadays, the aircrafts for commercial industrial purposes have a significant percentage of composites (40–50%) used in their structures. Nonetheless, most of these

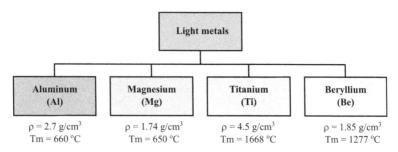

Fig. 1.1 Main light metals used as base materials in structural applications

composites are polymeric matrix composites, which have the important advantage of being easily manufactured, compared with metallic matrix composites. However, the mechanical properties and other important aspects of metal matrix composites can be better, for instance, when the component needs to be exposed to temperature during its service; in addition, metal matrix composites can be reused and recycled.

Conventional light metallic alloys, such as those made from aluminum, magnesium, and titanium, are still the most used ones in aircraft structures, which is due to their high specific properties; in the case of aluminum and magnesium, it is also due to their low processing temperatures (Fig. 1.1). Beryllium is also considered a light metal, and its alloys are being applied in aerospace and nuclear industries, such as in gyroscopes, inertial guidance systems, satellite antenna booms, and space telescopes; it is planned to use beryllium as a fuel element in power reactors [3].

1.1 Aluminum

The attraction to aluminum alloys can be attributed to their lightness, versatility of product shapes, high corrosion resistance, ease of maintenance, good appearance, and low-cycle costs, but also to high structural efficiency (large strength to specific weight ratio). Low density, a broad strength spectrum ranging from 70 to 800 MPa, nontoxicity, high thermal conductivity, high electrical conductivity, and a wide range of forming and working processes are just a few reasons why aluminum alloys are currently used [4].

Aluminum alloys have been the main airframe material for a century. Even though the role of aluminum in future commercial aircraft trends to decrease due to the increasing use of other composite materials, high strength aluminum alloys are still being used in aircraft structures. The aluminum alloys more used in the aeronautical industry are aluminum-copper (2XXX series) and aluminum-zinc (7XXX series). For example, 2XXX alloys are used in the manufacture of the lower wing skins and the fuselage structure of commercial aircrafts, while 7XXX alloys are used where a higher strength is required, such as in the upper wing skins. Some heat treatments and the reduction of impurities, in particular iron and silicon, have resulted in a higher fracture toughness and better resistance to fatigue crack initiation and growth. Taking advantage of their resistance-weight ratio, the aluminum alloys are widely used for manufacturing metal matrix composites. Unconventional manufacturing technology for aluminum alloys include high-speed machining and friction stir welding. High-speed machining allows the design of weight competitive high-speed machined assemblies. Friction stir welding is a new solid-state joining process, which has the ability to weld by fusion the difficult, or impossible, 2XXX and 7XXX alloys with less distortion.

Recently, aluminum-lithium (Al-Li) alloys have attracted the attention of researchers from both academia and industry sectors owing to Li, the lightest metal, can reduce the density of Al and increase its elastic modulus [5]. The increasing weight-to-strength ratio of Al-Li alloys opens the possibility of using them in aerospace structural applications.

1.2 Magnesium

Magnesium alloys are some of the lightest metal materials known, which compete with aluminum alloys for structural applications. However, magnesium alloys are not normally as strong as those of aluminum, having a lower modulus of elasticity. The biggest obstacle to the use of magnesium alloys is their extremely poor corrosion resistance. The most used magnesium alloys in the aeronautical industry are the magnesium-aluminum-zinc (AZXX) alloys, which have good mechanical properties despite their limited ductility. There exists the opportunity to improve the mechanical behavior of magnesium composites by using unconventional techniques, some of which will be covered in this book.

1.3 Titanium

Titanium alloys are one of the more important materials used in aircraft structures owing to their resistance to fatigue, high-temperature capability, and resistance to corrosion. In commercial aircrafts, the Boeing 747–100 contains only 2.6% of titanium alloys, while the newer Boeing 777 contains 8.3% [6]. Alpha-beta Ti-6Al-4 V alloy is still the most widely used titanium alloy, for which the melting practices have been improved, such as the multiple vacuum arc melting. Cold hearth melting is another new melting practice that combined with vacuum melting has proven to be essentially free of melt-related inclusions. In metal matrix composites, titanium alloys are very attractive because of their mechanical properties, but their manufacturing has many challenges that will be discussed in this book.

1.4 Beryllium

Beryllium alloys are also lightweight materials with an attractive combination of properties. However, beryllium must be processed using powder metallurgy technology and hot isostatic pressing. The latter is costly, and the process needs good control environments that increase the final cost. In addition, beryllium alloys have low thermal expansion, which provides them resistance to shape change when they are exposed to extreme temperature. This combination of characteristics makes beryllium ideal for demanding applications, including satellites and space structures. However, due to the high cost and little usage of beryllium, it will not be covered in this book.

1.5 Metal Matrix Composites

Metal matrix composites (MMCs) are materials which have at least two components: a matrix that is commonly made of a light alloy, such as aluminum, magnesium, and titanium, and the reinforcement, which can be any type of material usually stiff and strong, such as fibers like boron, silicon carbide and carbon, or ceramic particles. The main pursued specific properties are strength, stiffness, and toughness at elevated temperatures, and also creep and thermal shock resistance are desired behaviors. In recent decades, the use of nanoreinforcements in the synthesis of advanced materials has been widespread in many investigations. Carbon nanotubes and graphene are currently the most promising materials. Although low amounts of these reinforcements are needed (typically less than 2 wt%), their homogeneous dispersion within a matrix is maybe the most important and difficult challenge in the manufacture of composites. This is due to their very small dimensions and large specific surface area up to 200 m^2/g. Therefore, they tend to agglomerate and form clusters owing to van der Waals forces. A good dispersion of the reinforcement is needed for the efficient use of the properties, since nanoreinforcement clusters lead to a lower strength and higher porosity and serve as discontinuities, which decrease the mechanical, electrical, or thermal properties of the composite. Recent developments in this field seek for processes that promote a good dispersion of the reinforcement in the matrix with no damage, bringing at the same time an effective load transfer between the reinforcement and the metal matrix.

1.6 Reinforcing Materials

Reinforcing materials in MMCs are discrete fibers or second phase additions to a metallic matrix that result in an improvement of their mechanical properties, such as elastic modulus and strength. The most common reinforcing materials for MMCs are ceramics (oxides, carbides, nitrides, etc.), which are characterized by their high strength and stiffness. Reinforcements can be divided into two groups: (a) particulates or whiskers and (b) fibers (continuous and discontinuous). Fibers enhance strength in the direction of their orientation. In the case of continuous fiber reinforced MMCs, a lower strength in the direction perpendicular to the fiber orientation is obtained [7]. On the other hand, discontinuously reinforced MMCs exhibit more isotropic characteristics. Examples of common MMC reinforcements are [8]:

SiC Silicon carbide is by far the most important nonoxide ceramic reinforcement available commercially, being fibers and particles the two most common varieties. Another important type of SiC available for reinforcement purposes is whiskers.

Al_2O_3 Alumina can have different allotropic forms, namely, γ, δ, η, and α. The latter is the thermodynamically stable form. Many different alumina-based oxide fibers are available commercially. Additionally, particulate Al_2O_3 is also used for several engineering applications.

B Boron in fiber form has, like carbon fiber, high strength and high stiffness. It is commonly made by chemical vapor deposition (CVD) on a substrate such as tungsten or carbon. Boron on a tungsten substrate, denoted by B(W), finds applications mostly in aerospace and the sporting goods industries.

CNT Carbon nanotubes consist of concentric graphene cylinders produced in a low current furnace. They are produced as single-walled (SWCNTs) or multiwalled (MWCNTs) carbon nanotubes. CNTs have exceptional mechanical properties, which have motivated researchers to use them in the development of composite materials.

1.7 Production Techniques of Light Alloys and Composites

Liquid metallurgy allows manufacturing large quantities of materials by conventional casting equipment. However, in the case of composites, there are difficulties for achieving homogeneous dispersion of the reinforcement, poor wetting, and preferential formation of harmful interfacial products [3]. Among unconventional techniques for producing light alloys and composites (Fig. 1.2) are powder metallurgy combined with mechanical alloying and milling, sandwich technique, severe plastic deformation, additive manufacturing, and thermal spray coatings.

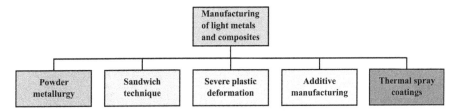

Fig. 1.2 Unconventional techniques for producing light alloys and composites

1.8 Powder Metallurgy

Powder metallurgy (PM) processing is the most economical and easy technique for manufacturing MMCs due to its simplicity and flexibility. The process involves mechanical blending of particles with metal powders in a mill, followed by other processes like compaction and sintering, cold isostatic pressing, hot pressing/hot isostatic pressing, or spark plasma sintering. These processes may be followed by secondary mechanical deformation treatments such as hot extrusion, hot forging, or hot rolling to consolidate the compacts into full-dense products. However, in some cases these processing techniques do not lead to a major increase of the mechanical properties due to the agglomeration of the reinforcement into the metal matrix. Most of these techniques involve high temperature, which can produce some damages to the reinforcement and compromise its stability during the fabrication of the composite. For this reason and as the materials are pushed near their limits, new matrices, reinforcements, and processes are demanded.

1.9 Mechanical Alloying and Milling

Initially developed by Benjamin [9], mechanical alloying (MA) and mechanical milling (MM) are techniques currently employed in the synthesis of advanced materials, being able of including equilibrium and non-equilibrium systems for the development of composites reinforced with a wide range of materials, such as carbides [10], oxides [11], as well as different types of fiber [12]. The technique is a solid-state powder processing that includes numerous events of cold welding, fracturing, and re-welding of powder particles in a high-energy ball mill. The advantage of MA/MM over standard metallurgy techniques is that they provide unique processing conditions in which solid-state reactions take place [13]. MA/MM avoids conventional melting and casting routes for fabricating alloys and composites, impossible to be produced by these standard practices [14, 15]. MMCs fabricated through MA/MM exhibit interesting and isotropic properties and attract considerable attention due to their relatively low cost [14].

Powder metallurgy and mechanical alloying are used as combined techniques in the fabrication of nanostructured materials with a reduced particle size and a diversity of structures [15], including some intermetallic and oxides [16–19] and carbon nanotubes [20]. The effect of the reinforcements is maximized when their size is reduced, as a result of the milling conditions [21, 22].

1.10 Sandwich Technique

The sandwich technique for the synthesis of metal matrix composites consists of staking alternate layers of reinforcement and metal like a sandwich structure and then consolidating by applying severe pressure [23]. Li et al. [24] have arranged 20 layers of 10-μm Cu foil with alternate CNTs layers of 450 nm in thickness and cold rolled the assembly with intermittent 24 annealing steps to form a Cu-CNTs composite. Yang et al. [25] studied the mechanical properties of Al/SiC nanolaminates with layer thicknesses between 10 and 100 nm. One of the promising techniques which has been developed is the sandwich technique [26]. This technique produces a material composed of a metallic matrix and banded structured layers of magnesium or aluminum reinforced with MWCNTs. The process consists of two steps: first pre-dispersion and pre-alignment of the reinforcement in a polymeric matrix and, second, staking metal and polymeric matrix composites consolidating by applying severe pressure. This technique also allows obtaining an increase in stiffness, strength, and fracture toughness compared with the matrix.

1.11 Severe Plastic Deformation

The techniques of severe plastic deformation (SPD) are experimental processes of metals forming where high plastic deformations are applied on a bulk process [27]. Metals of ultrafine grain (100–300 nm) can be obtained with SPD, having significant effects on their mechanical and functional properties [28]. The aim of these techniques is to produce lightweight materials of high mechanical strength [27]. Some examples of these techniques are equal-channel angular pressing (ECAP), accumulative roll bonding process (ARB), repetitive corrugation and straightening (RCS), multi-directional forging (MDF), twist extrusion (TE), cyclic extrusion and compression (CEC), and high-pressure torsion (HPT). In this book, the HPT technique, applied to titanium and magnesium alloys, will be presented as an effective route for the processing of light alloys and composites.

1.12 Additive Manufacturing

Additive manufacturing (AM) is based on a layer-upon-layer method to manufacture parts with complex geometries and for remanufacturing operations [29]. It was initially developed for the manufacturing of prototypes. However, in recent years the technique is being used in the design and manufacturing of tools with intricate design, such as cooling channels in injection molds, or medical and aerospace applications, among others [30, 31], as well as for producing functional parts without design constraints and constantly evolving with new applications [32, 33]. Even though these technologies focus on 3D printing for buildup structures, it must be mentioned that methods based on powder and wire-directed energy deposition are widely used on manufacturing and repair operations [34]. For the processing of MMCs by AM, powder bed, powder-directed energy deposition, and wire-directed energy deposition are the most promising techniques.

1.13 Thermal Spray Coatings

Thermal spray encompasses a group of processes in which metallic and nonmetallic materials and composites are sprayed as fine particles in a molten or semi-molten condition or even in fully solid state to form a deposited coating from flattened and solidified particles [35]. In the process, a heat source, which can be a combustion flame, an electric arc, or plasma, is used to melt the feedstock material (powder, wire, stick, suspension/solution) and accelerate the particles toward the solid surface of the substrate, which is the material or component to be coated. Other thermal or kinetic energy sources used are, for instance, laser beams in laser coating and inert gas flow jets in the cold spray process [36]. Materials suitable for thermal spraying cover practically all materials, including pure metals, metal alloys, hard metals (carbides), oxide ceramics, plastics, cermets, and blended materials. Compounds that decompose irreversibly during melting and the ones that do not have a stable melt state and vaporize excessively in the spray process are excluded. Depending on the applications, thermally sprayed coatings have thickness ranging from some micrometers up to several millimeters. Due to the versatility of thermal spray technique related with the use of extensive coating materials, widespread coating properties, high deposition rate, reliable, and cost-efficient, it is a widely used process in industry. For instance, thermal spraying to produce MMCs using nanoparticles or nanostructures as reinforcement phase has high relevance for technological application [37, 38] including aeronautical, automotive, and energy sectors, infrastructure, paper and printing industries, biomedical applications, electronics, etc., where properties such as friction, wear, temperature resistance, erosion, corrosion, adhesion strength, and surface finishing are some aspects of interest in research.

1.14 Characterization Techniques

Concerning the analysis of light alloys and composites, there are different characterization techniques available, ranging from the most common and accessible ones to the most sophisticated and, sometimes, not available to everyone. This book will inform the reader about the nuances and limitations of each technique. The composition of materials can be determined by chemical analysis in its different techniques like inductively coupled plasma (ICP) analysis. Thermal analysis is used to quantify the change in the material's properties with change in temperature; the most common techniques are differential scanning calorimetry (DSC), differential thermal analysis (DTA), and thermogravimetric analysis (TGA). The density of light alloys and composites can be measured by Archimedes' method or pycnometry. For the microstructural analysis, there exist different microscopy techniques, such as optical microscopy (MO), scanning electron microscopy (SEM), transmission electron microscopy (TEM) in its different modalities like high-resolution transmission electron microscopy (HRTEM), and scanning transmission electron microscopy (STEM). Energy-dispersive spectroscopy (EDS) and electron energy loss spectroscopy (EELS) are used to identify the chemical elements in the alloys and composites, as well as to identify the chemical interaction between the metal matrix and reinforcements. The structural study may be made by X-ray diffraction (XRD) and synchrotron high-energy X-ray diffraction. Atomic force microscopy (AFM) is used to study the morphology and to qualitatively identify the elastic modulus changes in the composites. X-ray photoelectron spectroscopy (XPS) is a technique used to study the surface of materials, providing quantitative and chemical state information from the surface of the material being studied. Raman spectroscopy is utilized to acquire information in a fast and global way; it is able to quantify, for instance, the content of a reinforcement present in a sample, their electronic nature, electric conductivity, and density of defects.

1.15 Mechanical Properties

In general, the mechanical properties of materials are related to the behavior under load or stress in tension, compression, or shear. The properties are determined by engineering tests under appropriate conditions. The mechanical characterization of light alloys and composites provides information about the strength, elastic limit, elastic modulus, elongation (ductility), creep strength, rupture strength, fatigue, impact strength (toughness and brittleness), and hardness. The mechanical properties are for determining in advance whether or not a material can be produced in the desired shape and can resist certain mechanical loads. In the case of composites, the mechanical properties depend on the chemistry, atomic structure, and the bonding at the interface, since this is the region where the mechanical load is transferred from the matrix to the reinforcement [39, 40].

The mechanical properties can be evaluated at different scales, going from the bulk scale, through the microscale and to the nanoscale. Most of the bulk scale tests are well established and standardized; tensile, compression, macrohardness, and tribological tests fall within this classification. The development of micro- and nanostructures requires new tools for their mechanical characterization, for which different testing techniques have been widely used to evaluate their mechanical properties at the small scale. Some of the challenges facing at this scale are the specimen preparation and manipulation and the high-resolution load and displacement sensing. Nanoindentation, nanoscratch, micro-cantilever beam, and atomic force microscopy (AFM) mapping are some of these small-scale tests.

1.16 Applications of Light Alloys and Composites

Nowadays, diverse light alloys with a great variety of features, properties, applications, and limitations are available. Selection of the type of alloy with characteristics that will depend on the expected application is a very important aspect from the point of view of its industrial implementation. Light alloys and their composites play an important role in the development of components for aeronautical, aerospace, defense, and automotive sectors, owing to their improved different properties and low weights. At present, for the proper selection of a material for any application, there are well-established tools and methodologies that can help minimize errors related to the material performance. However, the increase in productivity demands materials that impart better performances, which represent a challenge and opportunity to research. The manufacturing of better alloys is constantly related to the optimization and improvement not only of the material but also with the manufacturing processes. To the greatest extent, it is about controlling the composition minimizing the presence of defects due to impurities, inclusions, and shrinkage, to mention just a few. Improvements in material development represent a valid support for enhancing the life and reliability of an engineering component. Regarding the manufacturing processes, the efficient management of the productivity is very important for minimizing the cost and maintaining the high quality of the product. Even if the extensive production of light alloys considers conventional manufacturing processes, there is a general tendency to provide and develop innovative production processes, as well as new alloy compositions able to guarantee excellent performances, in a more profitable and timely manner [41]. These aspects are directly correlated to weight reduction and efficient fuel consumption, which constitute a central aspect for numerous industrial applications. In this book, an overview of the use of light alloys and composites for industrial components, as well as valuable research work conducted in the field by the authors with potential for industrial application, will be presented.

References

1. Peel, C., & Gregson, P. (1995). Design requirements for aerospace structural materials. In H. M. Flower (Ed.), *High performance materials in aerospace*. Dordrecht: Springer.
2. Ekvall, J., Rhodes, J., & Wald, G. (1982). Methodology for evaluating weight savings from basic material properties. In *Design of fatigue and fracture resistant structures*. Philadelphia: ASTM International.
3. Polmear, I., et al. (2017). *Light alloys: Metallurgy of the light metals*. Butterworth-Heinemann.
4. Dumitraschkewitz, P., et al. (2018). Clustering in age-hardenable aluminum alloys. *Advanced Engineering Materials, 20*(10), 1800255.
5. Prasad, N. E., Gokhale, A. A., & Wanhill, R. (2017). Aluminium–lithium alloys. In *Aerospace materials and material technologies*. Springer.
6. Williams, J. C., & Starke, E. A., Jr. (2003). Progress in structural materials for aerospace systems. *Acta Materialia, 51*(19), 5775–5799.
7. Faruk, O., Tjong, J., & Sain, M. (2017). *Lightweight and sustainable materials for automotive applications*. CRC Press.
8. Chawla, K. K. (2003). Ceramic matrix materials. In *Ceramic matrix composites*. Boston: Springer.
9. Benjamin, J. S. (1970). Dispersion strengthened superalloys by mechanical alloying. *Metallurgical Transactions, 1*(10), 2943–2951.

References

10. Clinktan, R., et al. (2019). Effect of boron carbide nano particles in CuSi4Zn14 silicone bronze nanocomposites on matrix powder surface morphology and structural evolution via mechanical alloying. *Ceramics International, 45*(3), 3492–3501.
11. Chen, C.-L., & Lin, C.-H. (2019). In-situ dispersed La oxides of Al6061 composites by mechanical alloying. *Journal of Alloys and Compounds, 775*, 1156–1163.
12. Suryanarayana, C. (2011). Synthesis of nanocomposites by mechanical alloying. *Journal of Alloys and Compounds, 509*, S229–S234.
13. Suryanarayana, C., Ivanov, E., & Boldyrev, V. (2001). The science and technology of mechanical alloying. *Materials Science and Engineering: A, 304*, 151–158.
14. Sundaresan, R., & Froes, F. (1987). Mechanical alloying. *JOM, 39*(8), 22–27.
15. Froes, F. (1990). The structural applications of mechanical alloying. *JOM Journal of the Minerals, Metals and Materials Society, 42*(12), 24–25.
16. Mehrizi, M. Z., & Beygi, R. (2018). Direct synthesis of Ti3AlC2-Al2O3 nanocomposite by mechanical alloying. *Journal of Alloys and Compounds, 740*, 118–123.
17. Luo, X.-T., Yang, G.-J., & Li, C.-J. (2012). Preparation of cBNp/NiCrAl nanostructured composite powders by a step-fashion mechanical alloying process. *Powder Technology, 217*, 591–598.
18. Wang, J., et al. (2013). In situ synthesis of Ti2AlC–Al2O3/TiAl composite by vacuum sintering mechanically alloyed TiAl powder coated with CNTs. *Journal of Alloys and Compounds, 578*, 481–487.
19. Karak, S., et al. (2018). Development of nano-Y2O3 dispersed Zr alloys synthesized by mechanical alloying and consolidated by pulse plasma sintering. *Materials Characterization, 136*, 337–345.
20. Pérez-Bustamante, R., et al. (2017). The effect of heat treatment on microstructure evolution in artificially aged carbon nanotube/Al2024 composites synthesized by mechanical alloying. *Materials Characterization, 126*, 28–34.
21. Prosviryakov, A., Samoshina, M., & Popov, V. (2012). Structure and properties of composite materials based on copper strengthened with diamond nanoparticles by mechanical alloying. *Metal Science and Heat Treatment, 54*(5–6), 298–302.
22. Prosviryakov, A. (2015). Mechanical alloying of aluminum alloy with nanodiamond particles. *Russian Journal of Non-Ferrous Metals, 56*(1), 92–96.
23. Salas, W., Alba-Baena, N., & Murr, L. (2007). Explosive shock-wave consolidation of aluminum powder/carbon nanotube aggregate mixtures: Optical and electron metallography. *Metallurgical and Materials Transactions A, 38*(12), 2928–2935.
24. Li, Y.-H., et al. (2007). Cu/single-walled carbon nanotube laminate composites fabricated by cold rolling and annealing. *Nanotechnology, 18*(20), 205607.
25. Yang, L., et al. (2016). Deformation mechanisms of ultra-thin Al layers in Al/SiC nanolaminates as a function of thickness and temperature. *Philosophical Magazine, 96*(32–34), 3336–3355.
26. Isaza Merino, C. A. (2017). *Study of the interface-interphase of a Mg-CNT composite made by an alternative sandwich technique*. Medellín: Universidad Nacional de Colombia–Sede Medellín.
27. Azushima, A., et al. (2008). Severe plastic deformation (SPD) processes for metals. *CIRP Annals, 57*(2), 716–735.
28. Valiev, R. Z., Zhilyaev, A. P., & Langdon, T. G. (2013). *Bulk nanostructured materials: Fundamentals and applications*. Wiley.
29. Milewski, J. O. (2017). Additive manufacturing metal, the art of the possible. In *Additive manufacturing of metals*. Springer: Cham.
30. Gonzalez, J., et al. (2019). Characterization of Inconel 625 fabricated using powder-bed-based additive manufacturing technologies. *Journal of Materials Processing Technology, 264*, 200–210.
31. Kumar, L., & Nair, C. K. (2017). Current trends of additive manufacturing in the aerospace industry. In D. Wimpenny, P. Pandey, & L. J. Kumar (Eds.), *Advances in 3D printing & additive manufacturing technologies*. Singapore: Springer.

32. Petrovic, V., Vicente Haro Gonzalez, J., Jorda Ferrando, O., Delgado Gordillo, J., Ramon Blasco Puchades, J., & Portoles Grinan, L. (2011). Additive layered manufacturing: Sectors of industrial application shown through case studies. *International Journal of Production Research, 49*(4), 1061–1079.
33. Pérez-Sánchez, A., et al. (2018). Fatigue behaviour and equivalent diameter of single Ti-6Al-4V struts fabricated by Electron Beam Melting orientated to porous lattice structures. *Materials & Design, 155*, 106–115.
34. Um, J., et al. (2017). STEP-NC compliant process planning of additive manufacturing: Remanufacturing. *The International Journal of Advanced Manufacturing Technology, 88*(5–8), 1215–1230.
35. Berndt, C. C., & Lenling, W. J. (2004). *Handbook of thermal spray technology*, ed. J.R. Davis. USA: ASM international.
36. Vuoristo, P. (2014) *Thermal spray coating processes*, in *Comprehensive materials processing*, ed. D. Cameron. Elsevier.
37. Bakshi, S. R., et al. (2009). Aluminum composite reinforced with multiwalled carbon nanotubes from plasma spraying of spray dried powders. *Surface and Coatings Technology, 203*(10–11), 1544–1554.
38. Yin, S., et al. (2018). Cold-sprayed metal coatings with nanostructure. *Advances in Materials Science and Engineering, 2018*, 1–19.
39. Schwartz, M. M. (1997). *Composite materials: processing, fabrication, and applications* (Vol. 2). Prentice Hall.
40. Desai, A., & Haque, M. (2005). Mechanics of the interface for carbon nanotube–polymer composites. *Thin-Walled Structures, 43*(11), 1787–1803.
41. Peter, I., & Rosso, M. (2015). Light alloys-From traditional to innovative technologies. In Z. Ahmad (Ed.), *New trends in alloy development, characterization and application*. IntechOpen.

Chapter 2
Manufacturing Processes of Light Metals and Composites

Abstract A variety of light alloys has been widely used from ancient to modern times. Aluminum is a light metal that is easy to melt, cast, and then process in a large variety of fabrication and forming processes. Being the lightest structural element, magnesium has become a premium choice in transportation industries. Most of these materials have been developed relying on many trial-and-error experiments, as well as the experience of researchers and companies. These alloys are needed for the development of new manufacturing routes even for the progress of improved alloys and nanocomposites. In the present age, computer-aided alloy design is, however, a useful tool to save time and cost necessary for the alloy development. This chapter presents a brief review of the manufacturing processes and some useful tools for the design of light alloys, which are the basis for the subsequent manufacturing process of nanocomposites.

2.1 Introduction to Manufacturing Process

The key to the effective use of lightweight materials is to tailor the material and processes to achieve the functionality of parts and components on a specific application. For instance, a sheet is used to create simple two- or three-dimensional shapes such as hoods, deck lids, and body panels. More complex three-dimensional forms with reinforcing webs are more commonly created from castings. In the selection of the fabrication method, cost and functionality must be analyzed to create both an efficient and cost-effective solution.

Because aluminum has a relatively low melting point, it can adapt to a wide variety of manufacturing processes, besides that it has attractive properties such as low density, good mechanical properties, and high corrosion resistance. At present, aluminum is commonly used for the manufacturing of industrial components in both the wrought (sheet, extrusions, and forgings) and cast forms. Alloying is important as it reinforces the material through the metallurgical mechanisms of solid solution strengthening and precipitation hardening. Additionally, heat treatments may be performed at any stage of the manufacturing process, including post-forming [1]. At an experimental level, such as that shown in Fig. 2.1, the solidification of light alloys with affectation of the cooling rate and subsequent heat treatments can be

Fig. 2.1 Processing of light alloys: (**a**) high-temperature furnace with agitation system, (**b**) experimental setup to solidify samples at different cooling rates, (**c**) copper, and (**d**) sand molds

studied. Figure 2.2 shows different samples obtained from experimental setup to study the solidification of light alloys.

The morphology of microstructures during solidification in aluminum and magnesium alloys will depend on the alloy content and the casting process. Aluminum is used as both wrought and casting forms. The production of wrought aluminum alloys involves the use of continuous or semicontinuous casting processes, in which intermediate to high cooling rates and steep temperature gradients are implicated [2]. Casting aluminum components are produced by different processes, such as sand casting, precision sand casting, lost foam casting, investment casting, die casting, and high-pressure die casting [3]. The use of semipermanent or permanent molding, as well as a chilling system, is possible. All these possibilities in the casting processes induce a large number of cooling rates and temperature gradients [2]. The selection of the casting process involves other factors, such as tooling, consumables, alloy costs, labor costs, thermal processing, and postproduction operations (weld repair, hot isostatic processing, impregnation) [3]. All these factors contribute to the overall unit cost of an aluminum casting. In addition to this, a predicted scrap rate has to be incorporated in the overall unit cost.

Most commercial aluminum and magnesium alloys are primary eutectic alloys, implying that solidification starts by nucleation, followed by dendritic growth and

2.1 Introduction to Manufacturing Process

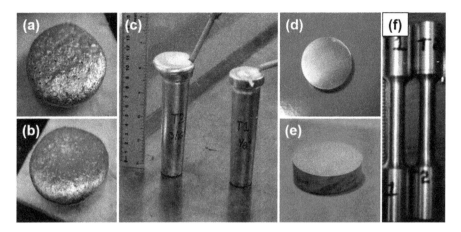

Fig. 2.2 (**a–b**) Samples for the evaluation of porosity. (**b**) Cast aluminum alloy extracted from metal molds. (**d–e**) Alloy sample for metallographic analysis. (**f**) Manufacturing of specimens for mechanical testing

ends by eutectic reactions. Figure 2.3 shows an example of experimental cooling curves obtained from solidifying a 319 aluminum alloy in sand and metallic molds; the latter were cooled using water. As can be seen, the solidification accelerates with the use of water-cooled metallic molds compared with typical sand mold. Short cooling times ensure a refined microstructure. An approach to early times shows the nucleation region of the solid through which the growth of the phases can be studied later.

Aluminum dendrites grow with orthogonal arms, which can be more or less developed depending on alloy composition and the cooling rate. As an example, Fig. 2.4 presents microstructures of 319 aluminum alloy samples processed under different conditions. As can be observed in Figs. 2.4c–d, the solution heat treatment causes aluminum dendrites to change in size, having an undeveloped arm. After the aging heat treatment (Figs. 2.4e–f), dendrites have the typical shape of well-refined wrought aluminum alloys containing only small additions of alloying elements, which segregate strongly promoting a fast dendritic arm growth. The morphology and distribution of phases in the overall aluminum matrix also change. For instance, Si phase (dark gray), coming from the eutectic reaction, is present in acicular geometry but fragmented.

In practice all commercial aluminum has refined grain, which is achieved through small additions (10–50 ppm) of certain elements like titanium and boron [2]. This results in a fine equiaxial microstructure, such as the example of Fig. 2.5, which shows the effect of the solution and aging heat treatments in the 319 aluminum alloy containing the addition of Ti and Ni (1 wt%) as grain refiner and reinforcement phase, respectively. The grain refinement mechanism is due to two effects: (i) the formation of TiB_2, which is insoluble in the aluminum melt, acting as a nucleating agent of aluminum crystals, and (ii) the presence of titanium in the liquid solution,

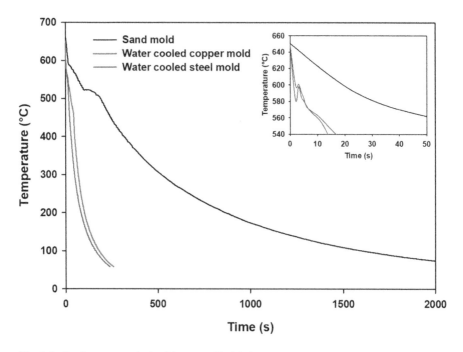

Fig. 2.3 Cooling curves obtained from modified 319 alloy cast into different molds

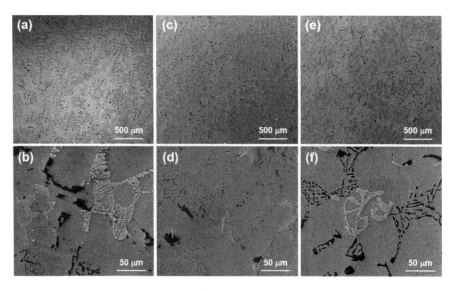

Fig. 2.4 Microstructures of cast 319 aluminum alloy under different conditions: (**a–b**) as cast, (**c–d**) solution heat treatment, and (**e–f**) aging heat treatment

2.1 Introduction to Manufacturing Process

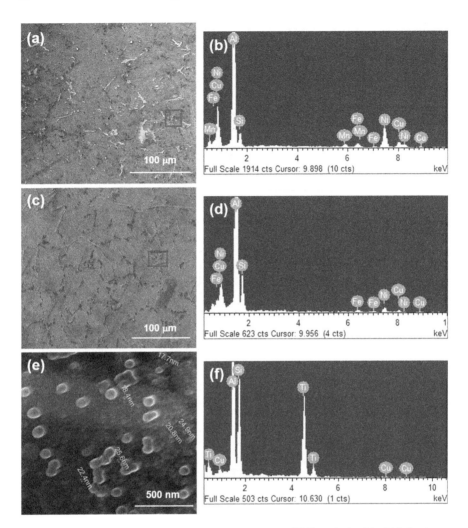

Fig. 2.5 Microstructures and energy-dispersive spectroscopy (EDS) analysis of the 319 aluminum alloy showing the effect of (**a–b**) the solution heat treatment, (**c–d**) the aging heat treatment, and (**e–f**) the Ti phase precipitated after aging

providing the necessary constitutional undercooling [4]. The grain refinement of aluminum provides better mechanical and surface properties and improves the alloy castability. The addition of Ni nanoparticles modifies the alloy microstructure, causing differences in the dendrite size and in the morphology of the present phases, with the consequent refinement of the light gray phase, which corresponds to α-AlFeMnSi with thin acicular morphology, and the dark phase corresponding to Al-Si phase. Figure 2.5e shows the Ti phase with a globular morphology, which precipitated after the aging heat treatment.

Magnesium is another lightweight material that shows promise in enabling significant reductions in vehicle weight. Its potential in both cast and wrought forms is being heavily investigated [5]. Similar to aluminum alloys, certain specific magnesium alloys can be heat treated by precipitation to strengthen the components. Some magnesium alloys rely upon solid solution strengthening for improving their mechanical properties. Magnesium alloys generally cast using high-pressure die casting processes and are not heat treated, even if they are likely to be. Structural components made from Mg are primarily limited to cast applications. This has been due to the relatively high cost of converting Mg into wrought products. Mg alloys manufactured by high-pressure die casting have moderate strength, high ductility, and good toughness, which make them attractive for components where crash response and energy absorption are important [1]. In aerospace industries, the use of magnesium is less than 1% of the aircraft structural mass, owing to (i) a limited toughness under static and dynamic loading due to a strong texture, (ii) high susceptibility to ignition, and (iii) poor corrosion resistance [6]. Although the poor corrosion resistance of magnesium alloys can be enhanced to an extent [7], their high flammability continues to remain a problem. Different modified processing techniques including hot extrusion [8], superplastic forming [9], twin rolling [10], and magnesium-based nanocomposites [11–13] have been studied to obtain desirable properties. Magnesium composites containing nanosized particulates are typically processed using similar methodologies as composites reinforced with microsized reinforcements, either by liquid-based or solid-based processing techniques.

The microstructure of magnesium can also be efficiently chemically grain refined. This can be done through zirconium additions and other elements compatible with zirconium including zinc, rare-earth elements, silver, and yttrium [5]. Nucleation of magnesium occurs by a peritectic reaction on a Mg-Zr intermetallic phase [2]. This sequence does not apply to magnesium alloys containing aluminum, owing to the formation of an Al-Zr intermetallic phase, which does not nucleate magnesium. Recent works in magnesium alloys, especially on strengthening effects of different alloying elements, depict the advances in strengths obtained [14–16]. Other works are not focused only in strengthening, rather in the effect of aluminum addition to Mg alloys on its formability [17], texture development [18], and corrosion properties as well [19, 20].

Figure 2.6 shows the Mg-rich part of the Mg-Al phase diagram with addition of 1.5 wt% Zn, which was made by the FactSage® software using the FTlite database [21]. Commercial alloys contain typically 2–10 wt% Al. Zn can be added up to 1 wt% and up to 3 wt% in alloys used as sacrificial anodes. Mn is always added to high-purity alloys to remove Fe. Mg and Al are fully soluble in the liquid state, and the eutectic reaction liquid → α (Mg) + β ($Mg_{17}Al_{12}$) takes place at 422 °C.

2.2 Unconventional Manufacturing Processes for Nanocomposites

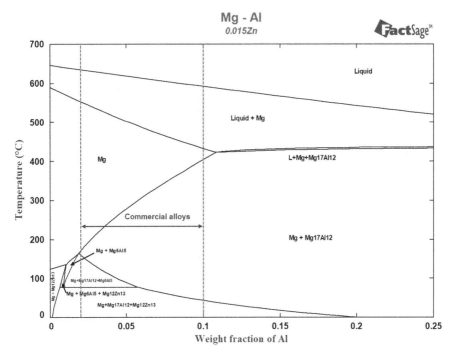

Fig. 2.6 Isopleth of the Mg-Al phase diagram with addition of 1.5 wt% of Zn

2.2 Unconventional Manufacturing Processes for Nanocomposites

Metal matrix composites (MMCs) are most promising materials in achieving enhanced mechanical properties such as hardness, strength, wear resistance, and fatigue resistance, among others. Aluminum-matrix composites reinforced with different nanoparticles are widely used for high-performance applications such as in automotive, military, and aerospace sectors because of their improved physical and mechanical properties [22]. The manufacturing methods of these composites can be classified into three types: (i) solid-state methods, (ii) semisolid-state methods, and (iii) liquid state methods. Liquid metallurgy technique is the most economical and effective of all the available routes for the production of MMCs, because it is a simple and inexpensive process, with flexibility and applicability to large-scale production, as well as excellent productivity for near-net-shaped components. However, the main limitation with this technique is the manipulation of the nanoscale with high temperature, which currently still represents an implementation challenge at the industrial level. Microsized particles usually improve the strength of materials they reinforce, but also significantly reduce their ductility, which limits their widespread application. The mechanical properties of materials would be further enhanced, while the ductility is retained by adding nanoparticles to the matrix.

2.2.1 Stir Casting

Metal treatment by means of stirring has been documented since the 1960s [3]. The applications are vast and have a wide range of purposes, such as degassing, grain refinement, mixing, desulfuring, removing inclusions, and iron passivation. A combination of good distribution and dispersion of microparticles can be achieved by stir casting [23, 24]. A uniform dispersion of submicron and nanoparticles can be achieved by ultrasonic-assisted stir casting; the process produces nonlinear effects such as cavitations, acoustic streaming, and radiation pressure, which give a uniform composite structure [25].

2.2.2 Semisolid Metal

In the semisolid metal (SSM) method, metals strength as a result of the dispersion of the non-metallic particles into the melt [26]. A condition for the method to operate successfully is that the microstructure of the starting material must consist of solid near-globular grains surrounded by a liquid matrix, as well as having a wide solidus-to-liquidus transition area. These microstructures impart thixotropic properties in slurries; that is, they have shear and time-dependent flow properties. SSM has found manufacturing applications on a number of applications due to its ability to deliver production of high-quality parts at costs that are comparable to or lower than conventional forming techniques, such as casting or forging. A unique property of the flow behavior in SSM is related to the non-Newtonian behavior of an alloy. When the material state is 50% solid and is sheared, the coalescence of the material will break up, its viscosity will fall, and it will flow like a liquid. If the material is allowed to stand for a certain time, the globular coalescence will increase its viscosity, which leads to it being able to support its own weight and be handled in the same way as if it was solid. When shear forces are applied, the near-globular particles move easily past one another, causing a decrease in viscosity and making the material behaves like a liquid. In contrast, when shear forces are applied on dendritic microstructures, typical of conventional castings, the liquid is trapped between dendrite arms and prevents them from moving freely, thus increasing the viscosity of the material. The SMM process has been studied in light alloys like Al7075 alloy [27], Al6Si3Cu alloy [28], and 2024 and 6082 wrought aluminum alloys [29].

2.2.3 Selective Laser Melting

Selective laser melting (SLM) is an additive manufacturing technology that offers significant potential for lightweight applications in space, aerospace, and automotive industries, as well as in mechanical engineering [30]. It is a commanding tool for the manufacture of metallic parts with a high degree of complexity. The process is characterized by the repeated creation of thin layers of metal powder and the selective laser scanning of successive cross sections of the parts being built. Laser radiation leads to full melting of the powder particles, and consequently almost fully dense metal parts can be built. Interesting fields include lightweight applications in space (e.g., brackets for satellites, spacecraft, and planetary rovers) and aerospace applications [31]. For lightweight applications, the structural potential of SLM can be further exploited by the use of aluminum alloys. To date, Al alloys most commonly processed by SLM are near-eutectic Si containing alloys, such as Al12Si and AlSi10Mg, which were originally developed for casting processes [32]. This technology offers benefits in terms of lightweight and functionally optimized parts and applications, especially in high-tech sectors. Thereby, the degree of an achievable mass reduction in lightweight applications depends not only on structural optimizations (e.g., the integration of lattice structures or topology optimization) but also on the mechanical properties of the lightweight material used, such as in Sc-modified aluminum alloys [32].

2.3 Design of New Alloys

The design of new alloys advances as the discovery of new materials progresses together with new processing routes and methodologies to perform their evaluation. Material discovery and design efforts ideally involve close coupling among materials prediction, synthesis, and characterization. The increased use of computational tools, the generation of materials databases, and advances in experimental methods have substantially accelerated these activities [33]. The design of new metallic alloys is faced with the challenge of an increasing complexity of the alloy composition, processing and resulting microstructures necessary to give a solution to multiple property targets, together with a requirement that the design stage be faster and less expensive [34]. Accelerated alloy design particularly challenging for structural materials is that their properties are not directly controlled by their composition but by their macro- and microstructure, among others at the nanoscale, which depends on both composition and processing parameters through phase transformations and their kinetics. The strategy for accelerated alloy design can be defined as a sequence of complementary objectives: (i) develop a knowledge-based capacity to predict the structure, microstructure, and properties of alloys, in order to minimize the number of experimental trials; (ii) develop high-throughput experimental and numerical methods allowing to explore increasingly large parts of the alloy design space, namely, chemical composition and processing parameters; and (iii) develop methods that allow finding an optimal alloy from the design space [34]. Evidently, these objectives are related, since the development and implementation of predictive models represent the basic stage of any optimization procedure and, alternatively, require large datasets to validate the model quality.

The development of advanced alloys is required to meet the increasing consumer demand for high-performance devices and components. The CALculation of PHAse Diagram (CALPHAD) methodology approach [35] has successfully assisted in designing advanced lightweight metallic materials, including those made from magnesium, titanium, aluminum, aluminum-based MMCs, and high entropy alloys, as recently summarized [36]. Due to various industry applications, many CALPHAD-based models have been proposed to describe the complex characteristics (thermodynamics, kinetics, strengthening, and thermophysical) in multicomponent systems [36].

The final performance of the desired materials is determined by the evolution of their microstructure and the formation of their phases. Therefore, the optimization of alloy compositions and processing routes via CALPHAD can provide significant information on phase formation and the properties of the desired alloys; thus, it may be useful to understand the alloy design. The main principle of this methodology is the minimization of the total Gibbs free energy (G) of an alloy system in a given condition of pressure (P) and temperature (T). The thermodynamic equilibrium of a system at constant pressure is given by the minimum of the Gibbs energy [35]. For multiphase equilibria, this means that the sum of the molar Gibbs energies for the stable phases is at a minimum, according to Eq. 2.1.

$$G = \sum_{\varphi} n^{\varphi} G_m^{\varphi} = \text{minimum} \qquad (2.1)$$

where n^{φ} is the number of moles and G_m^{φ} is the molar Gibbs energy of phase φ. The Gibbs energy of a phase can be written as the composition of different contributions, as stated in Eq. 2.2.

$$G_m^{\varphi} = {}^{ref}G_m^{\varphi} + {}^{cfg}G_m^{\varphi} + {}^{phy}G_m^{\varphi} + {}^{exs}G_m^{\varphi} \qquad (2.2)$$

where ${}^{ref}G_m^{\varphi}$ represents the frame of reference, such as the weighted molar Gibbs energies of the phase constituents or so-called endmember compounds, ${}^{cfg}G_m^{\varphi}$ is the configurational term, ${}^{phy}G_m^{\varphi}$ describes the contributions from other physical phenomena, such as magnetism, and the term ${}^{exs}G_m^{\varphi}$ is used to describe deviations of the Gibbs energy relative to the first three terms. Except for the configurational term, these terms are functions of temperature, pressure, and molar fractions of the constituents of the individual phases [35]. The constituents can be the elements, vacancies, or molecules in a disordered solution phase or so-called species in phases with an ordering tendency and can be elements, molecules, ions, or vacancies. Ordering tendencies are accounted for by describing a phase consisting of sublattices, which generally correspond to the different sites in the crystal structure of the phase. The molar Gibbs free energy (Gi) of each potential phase (i) is described as a function of its composition, P and T through mathematical expressions resulting from more or less complex theoretical and/or empirical thermodynamic calculations. Each term in Eq. 2.2 is adjusted to numerous measurements made on simple systems (mostly binaries, and ternaries) and/or with respect to ab initio simulations. Complex

2.3 Design of New Alloys

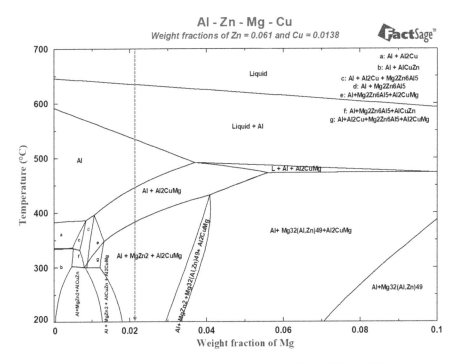

Fig. 2.7 Isopleth of the Al-Mg phase diagram with constant additions of Zn and Cu

numerical approaches, which are specific for each software, are then implemented to find the repartition of all elements in the alloy that gives the lowest possible value of G, i.e., the thermodynamic equilibrium. Possible outputs of computational thermodynamics are therefore the nature, fraction, and composition of phases but also thermodynamic quantities like activity, enthalpy, entropy, and free energy [34]. Commercial and open-source software suites now exist, along with thermodynamic databases for most categories of engineering alloys. CALPHAD methodology has been applied in the development of high-performance Al alloys. Recent case studies used this methodology to reasonably predict the solidification microstructure of the Al7Si2CuMg alloy, the complex solidification behavior of Al-(12–18)Si alloys and Al6Mg9Si10Cu10Zn3Ni alloys, and the precipitation behavior of the Al matrix [37]. Two case studies of the use of the CALPHAD methodology through the FactSage software [21] and FTlite database will be given below.

The first case study consisted of calculating the formation of phases and precipitates to optimize the aging heat treatment of an Al7075 alloy varying temperature (100, 120, and 150 °C) and time (8, 16, and 24 h). The chemical composition of the alloy was previously obtained by the inductively coupled plasma atomic emission spectroscopy technique and used as input into the program. The phases calculated were then validated through microstructure and crystallinity characterizations. Figure 2.7 shows the Al-Mg phase diagram with constant addition of Zn and Cu; the red line indicates the alloy chemical composition. Figure 2.8 presents the results of

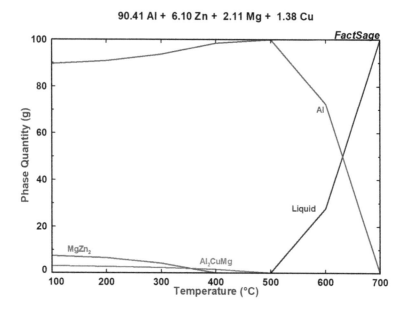

Fig. 2.8 Concentration profile in the Al7075 alloy

the concentration profile of main phases in the alloy as a function of the temperature. As can be seen, only two phases ($MgZn_2$ and Al_2CuMg) are preserved in low quantity in the alloy, according to the initial chemical composition of the aluminum matrix.

The phase formation was validated through X-ray diffraction (XRD) and transmission electron microscopy (TEM). As can be seen in Fig. 2.9, all heat-treated samples present the main XRD characteristic peaks of the α-Al (fcc) phase at $2\theta = 38.47°$ and $44.74°$, with no apparent displacement. In addition, peaks attributed to some secondary phases are observed: the Al_2CuMg phase at $2\theta = 38.40°$, $44.69°$, and $84.5°$ and the $MgZn_2$ phase, which precipitates during the aging heat treatment, at $2\theta = 37.11°$, $37.62°$, $40.06°$, $41°$, $41.73°$, and $43.31°$. The $MgZn_2$ phase is considered as metastable; hence it needs to be properly controlled during the heat treatment to reach its best dispersion into the aluminum matrix and contribute to an increase in the mechanical strength of the Al7075 alloy. Figure 2.10 presents TEM micrographs and an EDS elemental mapping obtained from the Al7075 alloy aged at 120 °C for 8 h. The precipitation of a nanostructured globular phase can be observed, whose chemical composition (Fig. 2.11) is very similar to that of the aluminum matrix, in addition to low quantities of Ti and Cr. The decrease in the aging time did not allow the complete diffusion of the solute atoms for the formation of the new precipitating phase $MgZn_2$, resulting in nano-precipitates with a composition similar to that of the aluminum matrix in the solid solution condition.

The second case study using the CALPHAD methodology consisted of investigating the thermodynamically equilibrium phases in a modified 319 aluminum alloy. The study was combined with the experimental characterization of a sample.

2.3 Design of New Alloys

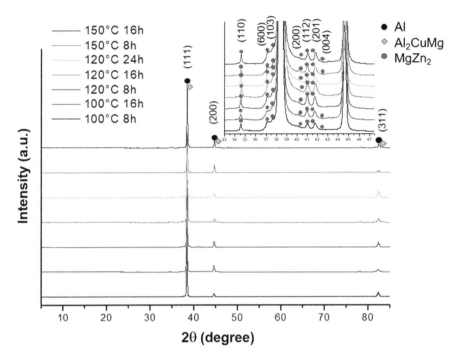

Fig. 2.9 XRD patterns of the Al7075 alloy aged under different conditions

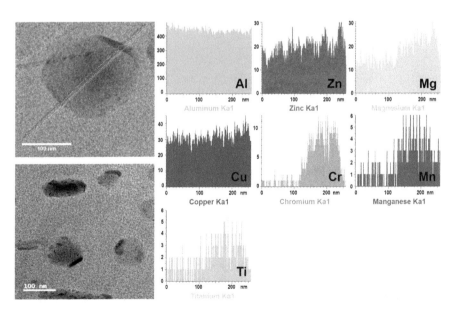

Fig. 2.10 TEM images and elemental mapping obtained from the Al7075 alloy after an aging treatment at 120 °C for 8 h

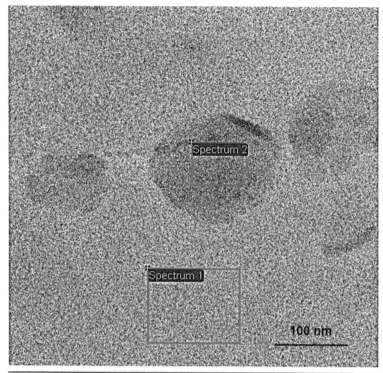

Spectrum	Chemical composition (wt%)							
	C	O	Mg	Al	Ti	Cr	Cu	Zn
1	7.98	1.80	2.34	82.91	-	-	2.94	2.03
2	7.73	2.42	4.55	77.99	0.43	1.22	3.31	2.36

Fig. 2.11 TEM image and EDS analysis obtained from the Al7075 alloy after an aging treatment at 120 °C for 8 h

Figure 2.12 shows a section of an isopleth of the Al-Si phase diagram with constant additions of Cu, Ce, Fe, Mn, and Ni. The quantities of each element are shown in the phase diagram, and the red line indicates the composition of interest (Al-7.8 wt% Si). As can be seen, this alloy is very complex, presenting six phases that could be present on the microstructure.

Figures 2.13 and 2.14 show the solidification profiles in equilibrium of the alloy, where the increase in the content of the phases formed as the temperature decreases can be observed. The points of liquidus and solidus of the alloys can also be observed in these concentration profiles. As can be seen in this alloy, the second most important phase apart from the aluminum matrix in terms of quantity (based on 100 g) is the silicon phase in almost 7.5 g, then the $CeCu_4Al_8$ phase in 4.3 g, and the α-AlFeMnSi phase in 2.5 g. Some phases in lower concentration, such as Al_3Ni and θ-Al_2Cu, are considered metastable compounds favored by the

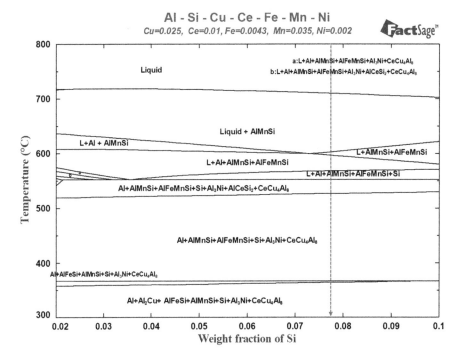

Fig. 2.12 Design of an Al-Si matrix-based light alloy with constant additions of Cu, Ce, Fe, Mn and Ni

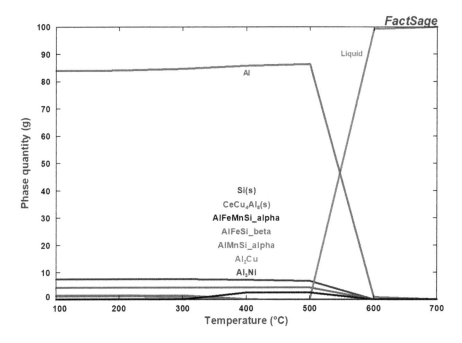

Fig. 2.13 Concentration profile of phases in the AlSiCuCeFeMnNi alloy

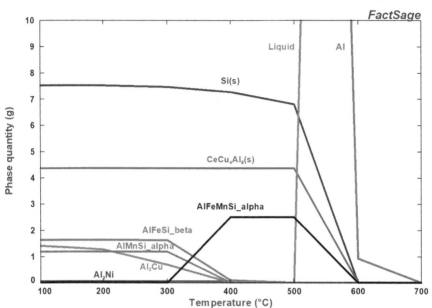

Fig. 2.14 Amplified concentration profile of phases in the AlSiCuCeFeMnNi alloy

non-equilibrium condition. These phases precipitate in the last stages of the solidification, even during the heat treatment, and can dissolve and reprecipitate, being able to enhance the mechanical properties of the alloy, depending on their distribution and morphology.

Concerning the experimental characterization, Figs. 2.15 and 2.16 present scanning electron microscopy (SEM) micrographs and energy-dispersive spectroscopy (EDS) analysis of the main phases in the aged alloy. As can be seen, the composition obtained by the EDS spectra corresponds to the phases identified by FactSage®. However, the micrographs reveal fragmented and rounded phases, compared to a conventional 319 alloy. The quantified elements in Fig. 2.15 match well with the Al, AlFeMnSi, Si, AlFeSi, and AlMnSi phases, while those quantified in Fig. 2.16 match well with the $CeCu_4Al_8$ and Al_3Ni phases.

Some unconventional techniques for the manufacture of recent light alloys and composites will be discussed in the subsequent chapters. Powder metallurgy, sandwich technique, severe plastic deformation, additive manufacturing, and thermal spray techniques have high potential to be implemented in the industrial sectors.

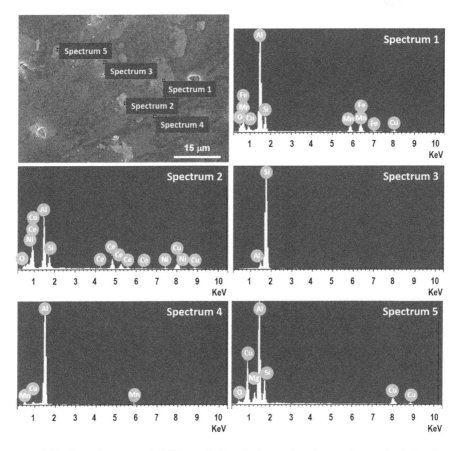

Fig. 2.15 SEM images and EDS analysis of the main phases characterized in the AlSiCuCeFeMnNi alloy

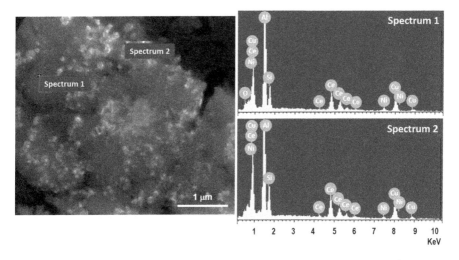

Fig. 2.16 Close-up of the spectrum 2 in Fig. 2.15, showing the phases containing Ce, Cu, and Ni

2.4 Conclusions

The review presented in this chapter dealt with the main processes and procedures for the manufacture of light alloys, comprising stages of melting, casting, solidification, and heat treatment, as well as novel techniques for the study and processing of nanocomposites. The use of computer-aided tools together with unconventional processes such as the selective laser melting process or ultrasonic-assisted stir casting represents methodologies for the study of new light alloys. Evidently, the experimental data are also useful to feed the databases, resulting in software predictions that deliver increasingly accurate results. The CALPHAD methodology, through FactSage® software, was applied in the development and discussion of high-performance Al alloys, using as case studies the hypoeutectic 7075 and 319 aluminum alloys. CALPHAD reasonably predicted the solidification microstructure, which was corroborated by SEM-EDS and TEM-EDS characterizations. In general, this methodology can be used to predict microstructures of magnesium, titanium, aluminum, and aluminum-based matrix composites, as well as high entropy alloys to accelerate the design and development of advanced lightweight materials. In the case of Al alloys, CALPHAD considers an alloying element content of 3–38 wt% and various casting processes including mold casting, direct-chill casting, and twin-roll casting. However, it is important to consider the possibility that the solidification microstructures slightly deviate from the simulation results due to solid-state diffusion occurring during the actual solidification and cooling processes. The CALPHAD methodology is useful for the design of new alloys and to predict precipitation behaviors in commercial Al alloys. This results in the optimizing of the alloy composition and processing routes to achieve the desired microstructure, thus improving the properties of light alloys and composites.

References

1. Kridli, G. T., Friedman, P. A., & Boileau, J. M. (2010). Manufacturing processes for light alloys. In *Materials, design and manufacturing for lightweight vehicles* (Woodhead Publishing Series in Composites Science and Engineering) (pp. 235–274). Dearborn: University of Michigan.
2. Arnberg, L. (2001). Solidification of light metals (Non-ferrous). In R. W. C. K. H. J. Buschow, M. C. Flemings, B. Ilschner, E. J. Kramer, S. Mahajan, & P. Veyssière (Eds.), *Encyclopedia of materials: science and technology*. Elsevier.
3. Hernandez Robles, F. C., Ramirez, J. M. H., & Mackay, R. (2017). *Al-Si alloys: Automotive, aeronautical, and aerospace applications*. Springer.
4. Easton, M. A., & Arvind Prasad, D. H. S. J. (2016). Grain refinement of aluminium alloys: recent developments in predicting the as-cast grain size of alloys refined by Al-Ti-B master alloys. In *Light metals*. TMS.
5. Westengen, H., & Rashed, H. M. M. A. (2016). Magnesium: Alloying. In *Reference module in materials science and materials engineering*. Elsevier.
6. Tekumalla, S., et al. (2018). A strong and deformable in-situ magnesium nanocomposite igniting above 1000 degrees C. *Scientific Reports, 8*(1), 7038.

References

7. Isaza Merino, C. A., et al. (2020). Mechanical and corrosion behavior of plasma electrolytic oxidation coatings on AZ31B Mg alloy reinforced with multiwalled carbon nanotubes. *Journal of Materials Engineering and Performance, 29*(2), 1135–1145.
8. Homma, T., Kunito, N., & Kamado, S. (2009). Fabrication of extraordinary high-strength magnesium alloy by hot extrusion. *Scripta Materialia, 61*(6), 644–647.
9. Kawasaki, M., & Langdon, T. G. (2015). Review: Achieving superplastic properties in ultrafine-grained materials at high temperatures. *Journal of Materials Science, 51*(1), 19–32.
10. Park, S. S., Park, W.-J., Kim, C. H., et al. (2009). The twin-roll casting of magnesium alloys. *JOM, 61*(8), 14–18.
11. Gupta, M., & Wong, W. L. E. (2015). Magnesium-based nanocomposites: Lightweight materials of the future. *Materials Characterization, 105*, 30–46.
12. Lan, J., Yang, Y., & Li, X. (2004). Microstructure and microhardness of SiC nanoparticles reinforced magnesium composites fabricated by ultrasonic method. *Materials Science and Engineering: A, 386*(1–2), 284–290.
13. Isaza Merino, C. A., et al. (2017). Metal matrix composites reinforced with carbon nanotubes by an alternative technique. *Journal of Alloys and Compounds, 707*, 257–263.
14. Huang, Z., et al. (2018). Observation of non-basal slip in Mg-Y by in situ three-dimensional X-ray diffraction. *Scripta Materialia, 143*, 44–48.
15. Zhang, J., et al. (2018). Recent developments in high-strength Mg-RE-based alloys: Focusing on Mg-Gd and Mg-Y systems. *Journal of Magnesium and Alloys, 6*(3), 277–291.
16. Luo, K., et al. (2019). Effect of Y and Gd content on the microstructure and mechanical properties of Mg–Y–RE alloys. *Journal of Magnesium and Alloys, 7*(2), 345–354.
17. Suh, B.-C., et al. (2014). Current issues in magnesium sheet alloys: Where do we go from here? *Scripta Materialia, 84-85*, 1–6.
18. Huang, X., et al. (2015). Influence of aluminum content on the texture and sheet formability of AM series magnesium alloys. *Materials Science and Engineering: A, 633*, 144–153.
19. Wang, X., et al. (2020). Corrosion behavior of as-cast Mg–5Sn based alloys with in additions in 3.5 wt% NaCl solution. *Corrosion Science, 164*, 108318.
20. Sun, Y., et al. (2020). Microstructure and corrosion behavior of as-homogenized Mg-xLi-3Al-2Zn-0.2Zr alloys (x = 5, 8, 11 wt%). *Materials Characterization, 159*, 110031.
21. *FactSage 7.3, CRCT-ThermFact Inc & GTT-Technologies*. 2006–2019; Available from: http://www.factsage.com/.
22. Koli, D. K., Agnihotri, G., & Purohit, R. (2015). Influence of ultrasonic assisted stir casting on mechanical properties of Al6061-nano Al2O3 composites. *Materials Today: Proceedings, 2*(4–5), 3017–3026.
23. Bhowmik, A., et al. (2020). Investigation on wear behaviour of Al7075-SiC metal matrix composites prepared by stir casting. *Materials Today: Proceedings*.
24. Chaubey, A., et al. (2020). Experimental inspection of aluminium matrix composites reinforced with SiC particles fabricated through ultrasonic assisted stir casting process. *Materials Today: Proceedings*.
25. Mula, S., et al. (2009). On structure and mechanical properties of ultrasonically cast Al–2% Al2O3 nanocomposite. *Materials Research Bulletin, 44*(5), 1154–1160.
26. Mohammed, M. N., et al. (2013). Semisolid metal processing techniques for nondendritic feedstock production. *ScientificWorldJournal, 2013*, 752175.
27. Chayong, S., Atkinson, H. V., & Kapranos, P. (2005). Thixoforming 7075 aluminium alloys. *Materials Science and Engineering: A, 390*(1–2), 3–12.
28. Alhawari, K. S., et al. (2017). Effect of thixoforming on the wear properties of Al-Si-Cu aluminum alloy. *Journal Teknologi, 79*(5–2).
29. Curle, U. A. (2010). Semi-solid near-net shape rheocasting of heat treatable wrought aluminum alloys. *Transactions of Nonferrous Metals Society of China, 20*(9), 1719–1724.
30. Spierings, A. B., et al. (2017). Microstructural features of Sc- and Zr-modified Al-Mg alloys processed by selective laser melting. *Materials & Design, 115*, 52–63.

31. Türk, D. -A., et al. (2016). *Additive manufacturing with composites for integrated aircraft structures,* ed. S.f.t.A.o.M.a.P.E. (SAMPE). Long Beach, CA, USA.
32. Spierings, A. B., et al. (2016). Microstructure and mechanical properties of as-processed scandium-modified aluminium using selective laser melting. *CIRP Annals, 65*(1), 213–216.
33. Davydov, A. V., & Kattner, U. R. (2019). Predicting synthesizability. *Journal of Physics D: Applied Physics, 52*.
34. Deschamps, A., et al. (2018). Combinatorial approaches for the design of metallic alloys. *Comptes Rendus Physique, 19*(8), 737–754.
35. Kattner, U. R. (2016). The Calphad method and its role in material and process development. *Tecnol Metal Mater Min, 13*(1), 3–15.
36. Shi, R., & Luo, A. A. (2018). Applications of CALPHAD modeling and databases in advanced lightweight metallic materials. *Calphad, 62*, 1–17.
37. Jung, J.-G., et al. (2019). Designing the composition and processing route of aluminum alloys using CALPHAD: Case studies. *Calphad, 64*, 236–247.

Chapter 3
Powder Metallurgy

Abstract This chapter deals with the production of light alloys and composites by the powder metallurgy process. A description of the stages involved in the process is given. Light alloys and composites find in powder metallurgy a technology that allows the manufacture of parts with high precision in their production and minimum waste and where additional machining processes are not required for large batches in mass production. In combination with unconventional routes, powder metallurgy is able to synthesize advanced materials with attractive physical characteristics for the transport industry. The process is able to produce strengthened materials by the dispersion of nanometric reinforcing agents. The case study in this chapter is addressed to the use of carbon nanotubes as one of the reinforcing materials of high interest, due to their potential in developing new structural materials, such as aluminum-based composites.

3.1 Introduction

Powder metallurgy (PM) is a processing technique for the synthesis of alloys and composites. This chapter includes the study of conventional routes of powder metallurgy, as well as unconventional techniques based on powder manipulation through mechanical alloying (MA). The most common elements used in the production of light alloys through powder metallurgy processes and the various processing stages involved are described. Powder metallurgy-based processes are highly appreciated in the scientific community owing to their versatility in the production of advanced materials that are impossible to synthesize by liquid routes (Fig. 3.1) and whose current interests involve their manipulation by techniques based on additive manufacturing [1, 2]. The application of the powder metallurgy in the production of light alloys and composites involves additional processing techniques in its different stages. On the whole, the purpose is the development of a functional material, with potential applications in the various areas that involve the transportation industry [3].

Materials manufactured through powder metallurgy are constituted by metallic powders that can be produced by different means, ranging from chemical and physical methods to mechanical routes. Among them, atomization is the most common

Fig. 3.1 Schematic representation of a typical component produced via powder metallurgy

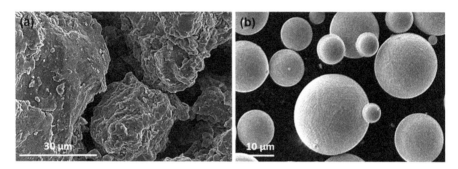

Fig. 3.2 SE-SEM micrographs of metal powder particles produced by (**a**) mechanical milling and (**b**) atomization (*Image courtesy: Cummins Inc.*)

and well-accepted technology employed for the powder production [4]. The characteristics of the powder production influence directly on their morphology. Atomization is the technology that allows to produce high-quality particles related to their spheroidization and uniform size [5]. Figure 3.2 shows images of powders produced by mechanical milling and atomization. Regardless of the production process, the application or use of metal powders in the industry involve various manufacturing and remanufacturing processes. These processes involve the formation of a solid mass in which the powders may or may not require a compaction process. Additive manufacturing is an example of a technique that does not require a compaction process, while in the sintering process, this stage is necessary. The sintering process can be carried out in conventional furnaces, induction or plasma systems, and laser systems [6, 7].

The powder metallurgy route, where a consolidation stage is involved for the production of functional parts, consists of a process that can be represented in the scheme shown in Fig. 3.3. The starting metallic powders and lubricant are compacted in a preset design die, reaching a state of consolidation commonly in green. The parts under this condition have low mechanical stability and require a sintering heat treatment to achieve the formation of metallurgical bonds among particles.

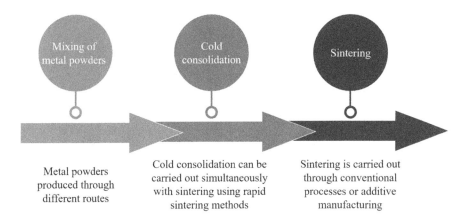

Fig. 3.3 Basic steps in the powder metallurgy process

3.2 Advantages of Powder Metallurgy

The versatility of powder metallurgy is such that it is currently used by various industries in the manufacture of parts that require a calculated control in the porosity, as well as high precision in the design, while simultaneously focusing on mass production and their uniformity. The complexity in the geometries and the accuracy in the control of the alloying elements that will constitute an alloy, which can be obtained by this process, are hardly attainable through conventional casting routes [8]. The powder metallurgy allows to minimize the machining process in those parts where the tolerance required in their application is a critical factor for their use and operation. This reduces in a complementary way the amount of waste material, making this technique an economically viable solution in the industry of the metal forming. These characteristics make powder metallurgy a sustainable technology for mass production of components with applications in various areas of the industrial sector [9].

3.3 Metal Powder Materials for the Development of Light Alloys

The starting materials used in powder metallurgy processes can be produced through different routes, such as grinding and atomization processes, among others. The mechanical route involves the constant fracture of materials until the particle size is reduced (Fig. 3.4). This process allows, in many cases, the reuse of materials that have been previously discarded from machining processes. The materials produced by this technique can be used in conventional powder metallurgy routes that involve cold consolidation and sintering [10], as well as some additive manufacturing

Fig. 3.4 (a) Secondary and (b) backscattered electron SEM micrographs of aluminum powders processed via mechanical milling

Fig. 3.5 Secondary electron SEM micrographs of the typical morphology of aluminum powders produced by air atomization. (*Images courtesy: Dr. Cynthia Deisy Gomez-Esparza*)

technologies such as cold spray. The morphology of particles that is obtained by mechanical means is directly related to the conditions and grinding equipment used in their processing [11, 12], obtaining as a result semispherical particles with a wide size distribution. This characteristic is apparently used to achieve bulk products, whose interstices can be occupied by smaller particles, obtaining as a final product a material with a high density.

On the other hand, most of the powder-shaped materials are produced through methods based on atomization systems, being water and gas the most commonly used for breaking up the stream of molten metal (Fig. 3.5). This process leads to the production of more uniform and edge-free particles in their morphology and whose quality can reach a high degree of spheroidization and batch uniformity. In the case of components made from particles with a high degree of spheroidization, the mechanical bonding among particles, product of the compaction stage, leaves a greater number of interstices among them and, therefore, a greater porosity [13].

3.4 Consolidation of Metal Powders

Regardless of the route of production of metal powders, the powder metallurgy process involves the mixing and compression of particles in a certain way, as well as the use of lubricants and binders. Figure 3.6 outlines the single cold compaction process. The powders are first loose with no strength and a large number of voids (Fig. 3.6a). As pressure is applied (Figs. 3.6b–c), the particles rearrange, and the contact among them increases. The increment in pressure is responsible for a better packing with the subsequent decrease in porosity, as well as improved intimacy among particles that increases the powder density (Fig. 3.6d). The applied pressure to form the cold compacts should be above the yield strength of the powders. The compaction process also serves to induce elastic and in some cases plastic deformation, which induce strain hardening of the compact and promote recrystallization that is critical for sintering. The consolidation step is conducted to allow proper handling of the samples prior to the sintering process. As an example of the cold compactions process, Fig. 3.7 shows the steps followed for the compaction of aluminum powders reinforced with carbon nanotubes.

The degree of compressibility is directly related to the theoretical density and depends on the nature of the material. In the case of aluminum and its alloys, which have a high degree of compressibility, it is possible to achieve a theoretical density greater than 90% at 165 MPa, while iron dust requires four times more pressure to reach a similar density value. Lubricants, such as calcium stearate in aluminum alloys, have the function of reducing the degree of abrasiveness between the

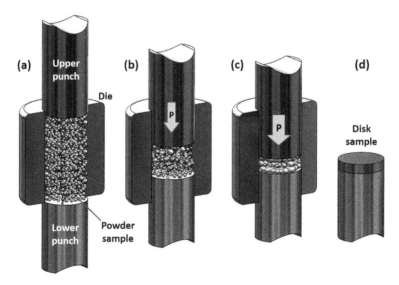

Fig. 3.6 Schematic representation of the single compaction process carried out during the powder metal processing. (**a**) The metal powder is placed in a die. (**b–c**) A load is applied to consolidate the loose powder. (**d**) A green product is obtained

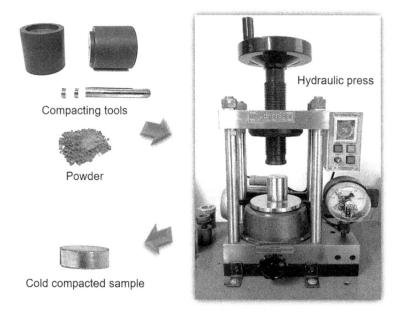

Fig. 3.7 Cold compaction processing of nanocomposite powders using a metallic die and a hydraulic press

material to be compacted and the walls of the tools, which are part of the compacting system [14]. The binder, mostly used in iron and its alloys, has the function of increasing the compaction resistance and giving stability in the green stage, while in aluminum alloys, it is rarely used due to their high compressibility.

3.5 Sintering of Metal Powders

The sintering process is a heat treatment that allows to transform the compacted powder into a consolidated sample, which occurs through thermal activation and interdiffusion processes, helping in reducing porosity and coarsening the powders [15]. These processes generally occur below the melting point of the metal particles, which forms the alloy or matrix of the composite material [16]. The sintering temperature is said to be approximately $T_{sint} = 0.8 T_M$, where T_M is the melting temperature in absolute scale [17]. With the application of temperature, the surface contact among the metal particles increases and consequently their strength. In some cases, chemical reactions occur during sintering; these chemical reactions are designed to enhance the bonding to form a coherent body. This process is accelerated with temperature. The resulting material has improved properties, and, in most cases, they are comparable or superior to the materials produced by casting. The sintering

3.5 Sintering of Metal Powders

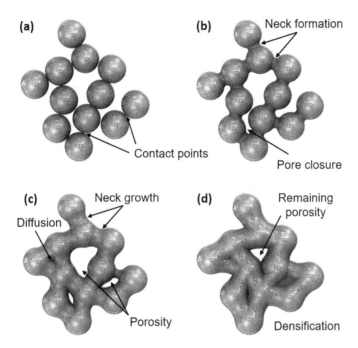

Fig. 3.8 Stages of the sintering process

process is carried out through several routes, ranging from different types of furnaces to laser systems.

Figure 3.8 presents the sequence of the sintering process. The particles have initial points of contact (Fig. 3.8a), which begin to be more bonded by the neck formation (Fig. 3.8b). A coarsening of the necks occurs over time (Fig. 3.8c) by mass transport, which is promoted by diffusion [15]. It is worth saying that some rapid sintering systems may be diffusionless [18, 19], but in conventional sintering, the driving force for sintering is diffusion. These transport phenomena are responsible for reducing most of the porosity, but there are always remaining pores among the sintered particles (Fig. 3.8d). The particle coarsening promotes coalescence. There are several aspects that lead to the final properties in the sintered products, which include temperatures and times. The temperature range for sintering is rather narrow for most alloys; however, lower temperatures are recommended as they tend to produce better sintered products, at the expense of longer times.

Furnaces operated by means of electric or gas resistances are used in conventional sintering processes. The pieces to process may be placed on conveyor belts for their handling in mass production processes. In the case of aluminum alloys, an inert atmosphere is required to prevent the formation of undesirable surface films, such as aluminum oxide [20]. This technique is the most used in the mass production of metal components, as well as in the study of experimental materials at the laboratory scale.

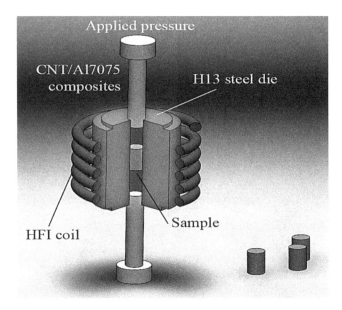

Fig. 3.9 Scheme of the high-frequency induction heating system employed in the sintering of metal powders

However, sintering processes consider additional technologies, where the use of high-frequency plasma or induction systems (HFIS, Fig. 3.9) is also explored because of the advantages offered by their high sintering speed rate [21], which is simultaneously carried out with the consolidation process. Unlike conventional sintering processes, which require long periods of heating and cooling, one of the advantages provided by rapid sintering methods is that they provide the amount of heat needed in the area of interest, in the period required by the product in its green stage. This is necessary to carry out the interdiffusion mechanism and the subsequent formation of metallurgical bonds among the metal particles. In the case of HFIS systems for sintering aluminum samples, the process can be carried out under atmospheric conditions to prevent the formation of surface oxide.

3.6 Powder Metallurgy and Mechanical Alloying in the Production of Light Alloys Strengthened with Carbon Nanotubes

As stated above, there is a great interest in the development of alloy and composites whose properties meet the energy saving requirements that modern industry demands. For this reason, PM, as one of the alternative technologies to foundry routes, allows the synthesis of advanced materials with a high degree of precision. These characteristics and benefits are employed in the synthesis and production of

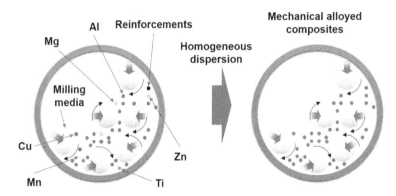

Fig. 3.10 Schematic representation of the mechanical alloying process for the synthesis of advanced materials

light alloys, especially those of aluminum. This has allowed the development of new materials and their strengthening with reinforcing agents with a scientific and technological interest. The automotive and aeronautical industries particularly find in aluminum-based materials a solution for the replacement of energy-efficient structural components when low weight and high strength are needed [22].

The interest in the use of aluminum alloys as structural materials was reinforced with the discovery and use of nanoparticles of diverse nature for the improvement of their physical properties. However, even when PM-based routes allow homogeneous mixtures of different alloying elements, the nanoparticle dispersion represents a challenge to overcome. Mechanical milling, in combination with PM techniques, offers a solution to the homogeneous dispersion of micro- and nanoparticles [23]. This technique is based on the processing of solid-state metal powders where repeated cycles of welding and particle fracture are involved in a given time until a homogeneous mixture between the reinforcing material and the matrix of the alloy or composite material is achieved (Fig. 3.10). This combination of techniques has been reported as a highly efficient route in the dispersion of reinforcing materials in various light alloys [24].

Carbon nanotubes (CNTs) are considered as a reinforcing agent in the production of advanced materials (Fig. 3.11) [25]. This is because they have a high elastic modulus and extreme tensile strength. Their mechanical properties and chemical stability at high temperatures make them an attractive material in the development of new applications. Their dispersion in light alloys, especially in aluminum alloys, has been achieved by combining the aforementioned techniques.

An efficient CNT dispersion in metal alloys can be carried out using high-energy systems, where the metal matrix can be constituted by commercial powders, or from metal chips obtained directly from machining processes. The reports in the literature are mainly those related with aerospace grade aluminum alloys [26, 27]. The reinforcing material, in this case CNTs, is synthesized by several routes, being methods based on chemical vapor deposition (CVD) the most efficient method for mass production [28]. In the production of Al-based composites, a reference sample

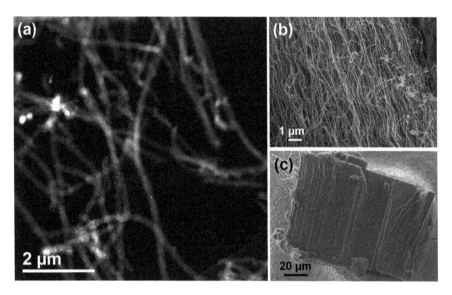

Fig. 3.11 (a) Z-contrast TEM micrograph of CNTs arrays. (b–c) Secondary electron SEM micrographs of CNTs

is always included and processed by the same route for comparative purposes. The MM process is carried out in inert conditions at atmospheric pressure using high-energy mills (Simoloyer, Spex, Emax). Process control agents, such as stearic acid or methanol, are included in the processing conditions to avoid excessive particle agglomeration and control their size throughout the milling time [29].

From this process, the powders obtained follow a consolidation process in which a conventional sintering system can be used at atmospheric pressure and inert atmosphere, or by means of rapid sintering systems, as has been the case with induction sintering at high frequency or plasma. When an aerospace grade forging alloy, e.g., 2XXX and 7XXX series aluminum alloys [25–27], is used as the basis of the composite, subsequent heat treatments are considered. In these kinds of alloys, a solubilization treatment followed by a natural or artificial aging process is carried out.

The mechanical characterization of the composites produced by this combination of techniques involves different mechanical tests, such as tension, compression, and microhardness tests, whose specimens are machined according to international standards. Another interesting concern in this area of study is the analysis of the microstructural behavior of the synthesized composites. The microstructural characterization considers the particle size obtained from the milling process and how it is affected by different aspects. Among these aspects, the following can be mentioned: doping elements; the degree of homogeneity in the dispersion of nanotubes achieved during the milling process, which is based on their concentration and milling time; and the formation of new phases resulting from the chemical interaction between the nanotubes and the aluminum matrix [30]. The analysis is carried out in the various stages that make up the combination of powder metallurgy and

3.6 Powder Metallurgy and Mechanical Alloying in the Production of Light Alloys...

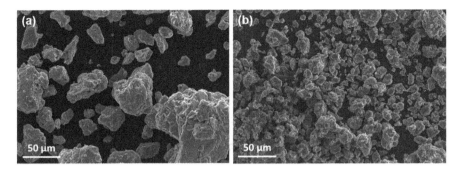

Fig. 3.12 Secondary electron SEM micrographs of as-milled powders. (**a**) Al7075 alloy and (**b**) Al7075–5.0 wt% CNT composite. The effect of CNTs as a grain refiner can be observed

mechanical milling techniques. This means that the powder analysis, obtained through studies that allow to establish the particle size distribution, and the laminar structure, obtained in the particle microstructure, are mainly determined by the milling time. The results derived from this analysis indicate a decrease in particle size inversely proportional to the CNT concentration [31]. Moreover, the increased milling time contributes additionally to the particle refinement. Although the effect of milling time has already been reported, the effect that CNTs produce on the particle size provides valuable information about their nature as a particle-refining agent (Fig. 3.12).

Although the formation of additional phases between aluminum and CNTs has not been reported at the stage of the powder synthesis, in the case of the sintering stage, comprehensive microstructural and X-ray diffraction analysis has indicated the formation of aluminum carbides in Al-CNTs composites, for CNT concentrations greater than 1.0 wt% [25, 31].

The analysis of the degree of dispersion of the CNT in the composite matrix is carried out mainly through electron microscopy studies [32]. For this purpose, microstructure analysis is performed in the cross section of selected samples, as well as fractography analysis (Fig. 3.13). The aim is to determine the type of fracture produced by the presence of a nanometric agent in the nature of the material [33]. Such dispersion analysis can be achieved in scanning electron microscopes, but on many occasions, it is necessary the use of transmission electron microscopes, where in addition to observing the dispersion of CNTs in micrometric areas it is possible to study the damage in the CNT structure caused by multiple impacts that occur during the mechanical milling process. Additionally, the formation of new phases, such as aluminum carbides, can be observed by this characterization technique.

The analysis by transmission electron microscopes involves the preparation of specimens to be studied by different techniques, such as ultramicrotomy, electropolishing techniques, or focused ion beam (FIB) systems. This last technique allows obtaining high-quality samples with a uniform thickness, with the disadvantage that the analysis area is limited to only a few tens of micrometers. Even so, such an area

Fig. 3.13 Secondary electron SEM fractography of CNTs embedded in an Al matrix

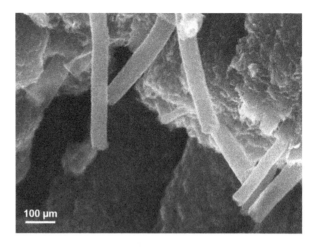

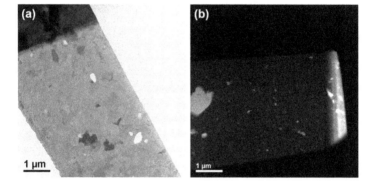

Fig. 3.14 (**a**) Bright field and (**b**) Z-contrast TEM micrographs of an Al7075-CNTs specimen prepared by FIB

provides a complete field of observation in which it is possible to study the interaction of the constituent phases of a material with its microstructure (Fig. 3.14).

The results obtained from the transmission electron microscope allow us to observe the dispersion of CNTs in the aluminum matrix, as well as their function as reinforcing phases. In relation to the presence and integrity of the CNTs, a degradation in their structure has been observed as a function of the milling time. Such degree of damage can be perceived as a loss of periodicity in the outer walls that constitutes their structure, whose origin is the multiple impacts caused during the mechanical milling process [34].

In those aluminum alloys that are susceptible to heat treatments, the presence of CNTs in aerospace grade aluminum alloys has shown that it does not alter the precipitation sequence, but its kinetics. The temperatures and times, in which the

3.6 Powder Metallurgy and Mechanical Alloying in the Production of Light Alloys... 45

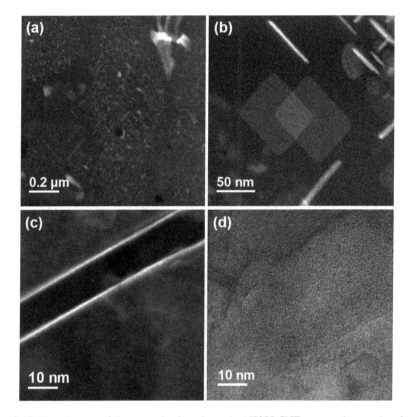

Fig. 3.15 TEM images of the strengthening phases in Al7075-CNTs composites produced via mechanical milling. (**a–b**) Cu-rich phases occurred during solubilization and artificial aging heat treatments. (**c**) Aluminum carbide formed during the chemical interaction between CNTs and the Al matrix. (**d**) A CNT embedded in the Al matrix

precipitation of phases is carried out after the solubilization treatment, as well as during the artificial aging heat treatment, depend on the concentration of CNTs (Fig. 3.15).

Concerning the hardening mechanisms, the loss of periodicity observed in the atomic arrangements, which constitute the CNT outer walls, suggests the formation of a CNT-Al interface, in which the load transfer from the matrix material to the reinforcing material is effectively carried out. Additionally, highly probable strengthening mechanisms have been considered due to the nanometric nature of the reinforcing materials used in the synthesis of these types of composites [35]. The inhibition in the movement of dislocations by the CNT presence and the thermal mismatch between the CNTs and the aluminum phase are mechanisms that pose interesting study paths aimed at a deeper understanding in the synthesis of Al-CNTs composites and their mechanical behavior.

3.7 Conclusions

Powder metallurgy continues being one of the most versatile techniques to produce near net shaped components with complex geometry and superior properties compared to those produced by conventional casting routes. The high-dimensional precision and uniformity in mass production systems make powder metallurgy a profitable manufacturing process in the transportation industry. The flexibility of this technique is such that, in combination with unconventional techniques, allows it to obtain materials with attractive physical characteristics and potential applications in the development of structural components. This multifaceted metal-forming process, jointly with mechanical alloying and mechanical milling, is considered a suitable technology that allows the use of diverse types of reinforcements, which makes the difference in achieving components with extraordinary properties demanded by the aeronautical and aerospace industries.

References

1. Chen, B., et al. (2020). Recent progress in laser additive manufacturing of aluminum matrix composites. *Current Opinion in Chemical Engineering, 28*, 28–35.
2. Ghods, S., et al. (2020). Electron beam additive manufacturing of Ti6Al4V: Evolution of powder morphology and part microstructure with powder reuse. *Materialia, 9*, 100631.
3. Rojas-Díaz, L. M., et al. (2020). Production and characterization of aluminum powder derived from mechanical saw chips and its processing through powder metallurgy. *Powder Technology, 360*, 301–311.
4. Garboczi, E. J., & Hrabe, N. (2020). Particle shape and size analysis for metal powders used for additive manufacturing: Technique description and application to two gas-atomized and plasma-atomized Ti64 powders. *Additive Manufacturing, 31*, 100965.
5. Vasquez, E., et al. (2020). Effect of powder characteristics on production of oxide dispersion strengthened Fe14Cr steel by laser powder bed fusion. *Powder Technology, 360*, 998–1005.
6. Lee, J.-H., et al. (2020). Constitutive behavior and microstructural evolution in Ti–Al–Si ternary alloys processed by mechanical milling and spark plasma sintering. *Journal of Materials Research and Technology, 9*, 2247.
7. Mendoza-Duarte, J. M., et al. (2020). Aluminum-lithium alloy prepared by a solid-state route applying an alternative fast sintering route based on induction heating. *Materials Letters, 263*, 127178.
8. Akhtar, S., et al. (2018). Recent advancements in powder metallurgy: A review. *Materials Today: Proceedings, 5*(9, Part 3), 18649–18655.
9. Vasanthakumar, P., Sekar, K., & Venkatesh, K. (2019). Recent developments in powder metallurgy based aluminium alloy composite for aerospace applications. *Materials Today: Proceedings, 18*, 5400–5409.
10. Rodríguez-Cabriales, G., et al. (2019). Synthesis and characterization of Al-Cu-Mg system reinforced with tungsten carbide through powder metallurgy. *Materials Today Communications*, 100758.
11. Abu-Warda, N., et al. (2018). The effect of TiB2 content on the properties of AA6005/TiB2 nanocomposites fabricated by mechanical alloying method. *Powder Technology, 328*, 235–244.

12. Singh, P., et al. (2019). Effect of milling time on powder characteristics and mechanical performance of Ti4wt%Al alloy. *Powder Technology, 342*, 275–287.
13. Polozov, I., et al. (2019). Synthesis of titanium orthorhombic alloy spherical powders by mechanical alloying and plasma spheroidization processes. *Materials Letters, 256*, 126615.
14. Kislyi, P. S., et al. (2010). Preparation of materials from aluminum oxide nanopowders using modern methods of consolidation. *Journal of Superhard Materials, 32*(6), 383–388.
15. Higashitani, K., Masuda, H., & Yoshida, H. (2006). *Powder technology: Fundamentals of particles, powder beds, and particle generation*. Boca Raton: CRC Press.
16. Padmavathi, C., Upadhyaya, A., & Agrawal, D. (2012). Effect of sintering temperature and heating mode on consolidation of Al–7Zn–2·5Mg–1Cu aluminum alloy. *Bulletin of Materials Science, 35*(5), 823–832.
17. Upadhyaya, G. S. (1997). *Powder metallurgy technology*. Cambridge International Science Publishing.
18. Omori, M. (2000). Sintering, consolidation, reaction and crystal growth by the spark plasma system (SPS). *Materials Science and Engineering: A, 287*(2), 183–188.
19. Munir, Z., Anselmi-Tamburini, U., & Ohyanagi, M. (2006). The effect of electric field and pressure on the synthesis and consolidation of materials: A review of the spark plasma sintering method. *Journal of Materials Science, 41*(3), 763–777.
20. Schaffer, G. B., et al. (2006). The effect of the atmosphere and the role of pore filling on the sintering of aluminium. *Acta Materialia, 54*(1), 131–138.
21. Dewidar, M. (2010). Microstructure and mechanical properties of biocompatible high density Ti–6Al–4V/W produced by high frequency induction heating sintering. *Materials & Design, 31*(8), 3964–3970.
22. Popov, V. A., et al. (2018). In situ synthesis of TiC nano-reinforcements in aluminum matrix composites during mechanical alloying. *Composites Part B: Engineering, 145*, 57–61.
23. Li, Z., W. Wang, And J.L Wang, Effects of TiB2 on microstructure of nano-grained Cu–Cr–TiB2 composite powders prepared by mechanical alloying. Advanced Powder Technology, 2014. 25(1): p. 415–422.
24. Cambronero, L. E. G., et al. (2003). Mechanical characterisation of AA7015 aluminium alloy reinforced with ceramics. *Journal of Materials Processing Technology, 143-144*, 378–383.
25. Pérez-Bustamante, R., et al. (2017). The effect of heat treatment on microstructure evolution in artificially aged carbon nanotube/Al2024 composites synthesized by mechanical alloying. *Materials Characterization, 126*, 28–34.
26. Deaquino-Lara, R., et al. (2011). Synthesis of aluminum alloy 7075-graphite composites by milling processes and hot extrusion. *Journal of Alloys and Compounds, 509*, S284–S289.
27. Sameezadeh, M., Emamy, M., & Farhangi, H. (2011). Effects of particulate reinforcement and heat treatment on the hardness and wear properties of AA 2024-MoSi2 nanocomposites. *Materials & Design, 32*(4), 2157–2164.
28. Herrera-Ramirez, J. M., Perez-Bustamante, R., & Aguilar-Elguezabal, A. (2019). An overview of the synthesis, characterization, and applications of carbon nanotubes. In Y. Srinivasarao et al. (Eds.), *Carbon-based nanofillers and their rubber nanocomposites*. Amsterdam: Elsevier.
29. Dasharath, S. M., & Arati, C. (2020). Effect of Zr on nanocrystalline Al-4.5 wt.% Cu alloy and its strengthening mechanisms prepared by mechanical alloying. *Materials Today: Proceedings, 20*, 140–144.
30. Esawi, A., & Morsi, K. (2007). Dispersion of carbon nanotubes (CNTs) in aluminum powder. *Composites Part A: Applied Science and Manufacturing, 38*(2), 646–650.
31. Pérez-Bustamante, R., et al. (2011). Characterization of Al2024-CNTs composites produced by mechanical alloying. *Powder Technology, 212*(3), 390–396.
32. He, T., et al. (2017). The use of cryogenic milling to prepare high performance Al2009 matrix composites with dispersive carbon nanotubes. *Materials & Design, 114*, 373–382.

33. Pérez-Bustamante, R., et al. (2009). Microstructural and mechanical characterization of Al–MWCNT composites produced by mechanical milling. *Materials Science and Engineering: A, 502*(1), 159–163.
34. Pérez-Bustamante, R., et al. (2013). Effect of milling time and CNT concentration on hardness of CNT/Al2024 composites produced by mechanical alloying. *Materials Characterization, 75*, 13–19.
35. George, R., et al. (2005). Strengthening in carbon nanotube/aluminium (CNT/Al) composites. *Scripta Materialia, 53*(10), 1159–1163.

Chapter 4
Sandwich Technique

Abstract This chapter deals with the processing technique denominated sandwich technique for the synthesis of light alloy composites. The technique is described in detail for manufacturing magnesium and aluminum alloys reinforced with multiwalled carbon nanotubes. The microstructure of the products obtained at each step of the technique is discussed. The microstructural and mechanical characterization of the final products is presented and discussed. The sandwich technique consists of stacking a polymer previously reinforced with carbon nanotubes, which is used as a vehicle to bring the reinforcement to the metal matrix. The manufacturing process is carried out by a hot compacted system with atmosphere and pressure control. The sandwich technique is a diffusive process where the polymer is thermally degraded, to finally clamp the reinforcement between metallic sheets. This chapter shows the feasibility of the manufacturing process for future applications in different sectors.

4.1 Introduction to the Sandwich Technique

The sandwich technique consists of stacking different layers of materials followed by compaction and rolling processes. This processing technique is normally used to manufacture fiber-reinforced metal matrix composites (MMCs) from sheets, foils, powder, powder tape, or wire as a matrix material. The methods for assembling reinforcing fibers and matrices in MMCs are diffusive processes. The composite elements are assembled by layering the fiber array and matrix plies to achieve a predetermined fiber orientation and composite thickness. Composite consolidation is achieved by applying a high pressure to the surfaces and temperature enough to produce atomic diffusion of the metallic matrix. However, a few processing techniques do not use temperature to produce atomic diffusion to manufacture layered composites but rather some adhesive processing, especially when dissimilar or incompatible materials are involved. Layered composites have been shown to have good mechanical properties such as elastic modulus, yield strength, ultimate strength, and fracture toughness.

Some researchers have synthesized MMCs reinforced with carbon nanotubes (CNTs) by staking alternate layers of CNTs and metal to form a sandwich structure, with the subsequent use of consolidation techniques. Salas et al. [1] explored the explosive consolidation for manufacturing alternate layers of Al powder and CNTs to produce composites with 2 and 5 vol% CNTs; a deterioration of the mechanical properties was observed due to the CNT clustering at the grain boundaries. Eizadjou et al. [2] produced multilayered composites by an accumulative roll-bonding process using Al 1100 and Cu strips. They observed that copper layers were necked and fractured after the rolling passes; homogeneously distributed fragmented copper layers in the aluminum matrix were evidenced. Further, the mechanical properties of these composites were increased by increasing the strain during the process. Garcia et al. [3] studied the interlaminar reinforcement using aligned CNTs for prepreg unidirectional carbon tape composites; note that in this case the material is a polymeric matrix composite. Yu et al. [4] used the layered structure for synthesizing hybrid composites, stacking aluminum, and CNT-reinforced epoxy followed by compaction; they found that the fracture toughness of their composites improved significantly with the CNT content.

As an example of the advantages of the sandwich technique in the synthesis of light alloys and their composites, this chapter will be focused on the production of AZ31B magnesium and 1100 aluminum alloys reinforced with multiwalled carbon nanotubes (MWCNTs). The technique consists of producing a material composed of a metallic matrix and banded structured layers of MWCNTs. This processing allows obtaining an increase in the mechanical properties of the composites, such as stiffness, strength, and fracture toughness, compared with those of the metallic matrix. In addition, compared to other investigations, the sandwich technique is promising for the technological application due to its low cost and versatility to process different light matrices and reinforcements.

4.2 Magnesium Alloys

Magnesium alloys are chemically active and can react with other metallic alloying elements to form intermetallic phases. These phases aid to influence the microstructure and, hence, affect the mechanical properties of the magnesium alloy and their composites. Aluminum and zinc are commonly used as alloying elements in magnesium alloys. The addition of aluminum results in the enhancement of hardness and strength, as well as of castability. Furthermore, aluminum makes magnesium alloys susceptible to heat treatment [5]. The addition of zinc is usually used together with aluminum to increase the strength without decreasing ductility. Zinc can also help to improve the corrosion resistance along with other elements, such as nickel and iron [5]. One of these kinds of alloys is the AZ31B alloy, which contains 3 wt% Al and 1 wt% Zn, among other elements. This alloy has good strength, ductility, corrosion resistance, and weldability at room temperature. Its applications are varied, ranging from automotive components, cell phone, and laptop cases, to aircraft fuselages.

4.3 Aluminum Alloys

Aluminum alloys have excellent corrosion resistance to natural atmospheres, suitability for food and beverage storage, high electrical and thermal conductivities, high reflectivity, and ease of recycling. Several Al alloys can be heat treated and loaded to a relatively high level of stress. Thanks to these properties, aluminum alloys have been used in a variety of applications, ranging from cooking utensils, car wheels, and engines, to aeronautical components. Despite the appearance of new materials such as composites, Al alloys still remain as important airframe structure materials because of their low cost, ease of manufacturing, and low density. Due to their superior damage tolerance and good resistance to fatigue crack growth, some Al alloys are used in fuselage skins, lower wing skins, and upper wing skins on commercial aircrafts [6]. However, aluminum alloys have some disadvantages, including low modulus of elasticity, low performance, and susceptibility to corrosion to elevated temperature. Due to this, aluminum matrix composites have become important alternative materials for aircraft applications. Accordingly, our goal in this chapter is to present the advantages of the alternative technique denominated sandwich technique in the synthesis of Al-based composites. The 1100 aluminum alloy, corresponding to the commercially pure aluminum, was chosen for this task. This alloy contains at least 99% aluminum with copper, iron, magnesium, manganese, silicon, titanium, vanadium, and zinc as remaining elements. It is soft, has low strength, can be cold worked extensively in the annealed condition without the need for intermediate annealing, and has excellent machinability. Its applications range from cooking utensils and decorative parts, to rivets, heat exchanger fins, and sheet metal works.

4.4 Metal Matrix Composites Synthesis

Metal matrix composite manufacturing is done in two steps. The first step is the pre-dispersion and pre-alignment of the MWCNTs within a polymeric matrix, to obtain a polymer matrix composite; in the present case, we used polyvinyl alcohol as a matrix. The second step is the manufacture of the metal matrix composites themselves by the sandwich technique.

4.4.1 Polymer Matrix Composites Synthesis

To maximize the advantage of nanoparticles and nanofibers as effective reinforcements in high strength composites, they should not form clusters and must be well dispersed into the matrix. To achieve these two aspects, there are several techniques to improve the dispersion of the nanoreinforcements in polymer matrices, such as ultrasonication [7, 8], high shear stress [9], calendering process [10], ball milling [11], stirring [12], and extrusion [13]. In fiber nanocomposites, both mechanical properties, such as stiffness and strength, and functional properties, such as electrical, magnetic, and optical properties, of polymer/nanoreinforcement nanocomposites are linked directly to the alignment of nanofiber in the matrix. Different methods for aligning nanofibers in polymeric matrices have been used, such as ex situ alignment [14], force field-induced [15], magnetic field-induced [16], electrospinning-induced [17], and mechanical stretching [8], among others. In this work, the ultrasonication dispersion method was used, which is described below.

The equipment setup used for the synthesis of the polymer matrix composites is shown in Fig. 4.1: (1) Pellets of fully hydrolyzed polyvinyl alcohol (PVA) were dissolved in hot distilled water to produce a solution of 4 wt% PVA. MWCNTs were introduced into the PVA solution in percentages of 0.25, 0.5, 1.0 and 2.0 wt%, where they were (2) dispersed by magnetic stirring during 1 h at an average speed

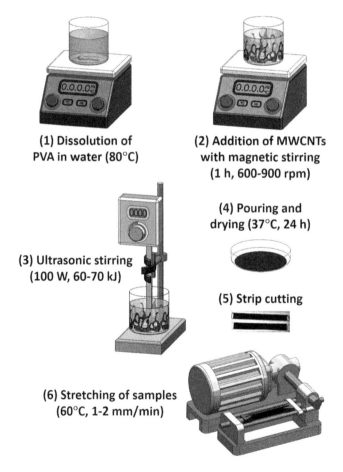

Fig. 4.1 Schematic representation of the PVA/MWCNTs nanocomposite synthesis

of 600–900 rpm. This was followed by (3) ultrasonication for 3 h at a power of 100 W and an amplitude of the probe of 20%; the dispersion maximum energy (60–70 kJ) was controlled to prevent damage to the CNTs [18, 19]. (4) The solution was poured and dried at 37 °C during 24 h for allowing the polymer to cure. (5) Strips were obtained and (6) mechanically stretched at a speed of 1–2 mm/min and temperature of 60 °C to obtain sheets with MWCNTs aligned in the stretching direction.

4.4.2 Metal Matrix Composite Synthesis

Two matrices were used for the synthesis of metal matrix composites: AZ31B magnesium alloy and 1100 aluminum alloy, which were reinforced with multiwalled carbon nanotubes.

Before carrying out the synthesis of the composites by the sandwich technique, sheets of AZ31B magnesium were annealed at 350 °C and then hot rolled at 250 °C until obtaining a thickness of about 150 μm; both processes were performed in environmental atmosphere, which allowed having a more homogeneous composite during the sandwich technique processing. The rolling process changed the microstructural morphology, as shown in Fig. 4.2. The microstructure of the as-received material evidences a heterogeneous grain size and some twins (Fig. 4.2a).

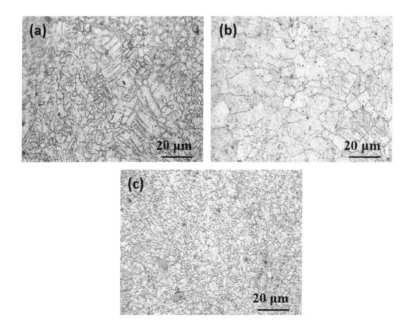

Fig. 4.2 Microstructural changes after recrystallization and deformation process for the AZ31B alloy: (**a**) as-received, (**b**) as-annealed, and (**c**) as-hot rolled conditions

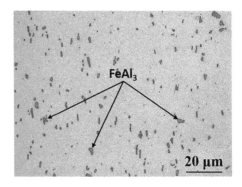

Fig. 4.3 Microstructure of the 1100 aluminum alloy after hot rolled processing

After the annealing process, the grain grows and twins disappear (Fig. 4.2b). Finally, the grain size decreases after the hot rolling process, while the twin formation is not clearly seen (Fig. 4.2c).

Similar processing was carried out for the 1100 aluminum alloy. As shown in Fig. 4.3, the microstructure of the Al 1100 alloy reveals the presence of the $FeAl_3$ phase, which is evidenced by small gray islands.

For the synthesis of the metal matrix composites, two sheets of the PVA/MWCNTs composites were alternately stacked with three sheets of either AZ31B alloy or 1100 alloy, as shown in the schematic images of Fig. 4.4. This stacking was hot compacted in a die under argon atmosphere to prevent the MWCNTs oxidation during the process. Thermogravimetric analysis [20] showed that when an argon atmosphere is used, the minimum temperature to totally evaporate the PVA and to leave the CNTs is about 500 °C. However, both the PVA evaporation and metal diffusion rates depend on time and temperature, which is why in this work the processing temperature was set near the magnesium and aluminum melting points, in order to increase both rates. Thereby, the temperature was gradually raised during 1.5 h up to 580 °C and 600 °C for magnesium and aluminum alloys, respectively; the pressure was also gradually raised up to 40 MPa. This was followed by a holding period of 1.5 h for allowing the PVA to evaporate and the metal matrix to diffuse among the sheets. Both kinds of composites were annealed during 30 min and finally hot rolled at 350 °C. The final product obtained were samples of approximately 300–500 μm in thickness. The zones located between metallic matrix sheets are termed "reinforced zones" in this work (Fig. 4.5), which were subjected to a close inspection. These results will be thoroughly discussed below.

4.4.3 Diffusion Bonding Mechanism

During the synthesis of metal matrix composites by the sandwich technique, the hot compaction stage allows the metal sheets to be joined by the diffusion mechanism. Diffusion bonding is the solid-state joining process by which two surfaces are joined under high temperature and interfacial pressure. This process is schematically rep-

4.4 Metal Matrix Composites Synthesis

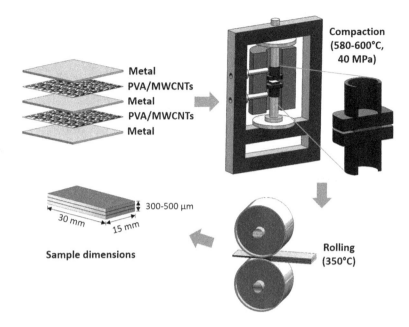

Fig. 4.4 Schematic representation of the metal matrix-PVA/MWCNTs composite synthesis

resented in Fig. 4.6: (a) The pressure application causes the surface asperities of both metals to come into contact. (b) The voids are reduced by plastic deformation. (c) The continuous yielding and creep lead to reduce voids. (d) Diffusion of material takes place due to the temperature effect; the vacancy diffusion allows to leave very few small voids. (e) The bonding is completed.

The diffusion bonding offers many advantages; for instance, the strength and microstructure of the bonding line are the same to the base metals. However, the behavior can be different with the presence of a reinforcement at the bonding line, as will be discussed later. On the other hand, the bonding process requires several

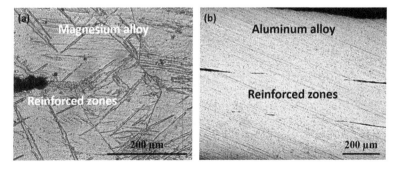

Fig. 4.5 Composite sections studied by optical microscopy: (**a**) AZ31B magnesium alloy and (**b**) 1100 aluminum alloy

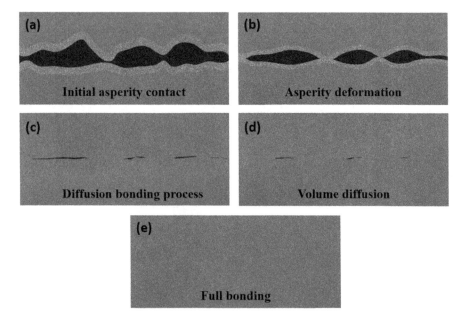

Fig. 4.6 The diffusion bonding process

strictly controlled conditions, such as cleaned and homogeneous surfaces of the metals involved. Further, in order to achieve a bonding strength, an appropriate pressure, time, and temperature must be selected, which depend on the physical and chemical characteristics of the materials.

The bonding aspect at elevated temperature changes the properties and microstructural behavior of the joined materials. To achieve an intimate contact between metallic cleaned surfaces, the diffusion process represents only a small part of the mass transport required for the joining, since most of it is achieved by plastic deformation during the compaction process. Due to the above, the low flow stresses associated with fine grain sizes are desirable for bonding, just as for superplastic forming. However, the diffusion bonding process can adversely affect the grain growth process; thereby it should be a matter of special concern for bonds to be achieved properly [21].

Several materials may be joined directly by diffusion bonding to form a solid-state welding. Similar materials, similar materials with a thin layer of a different metal between them, two dissimilar metals, and dissimilar metals may be joined with a third metal between them. The metallurgical factors associated with diffusion bonding are allotropic and microstructural transformations, which change the diffusion rates during the diffusion bonding process and the mechanical behavior of the joined materials. In the progress of these transformations, the metal has a high plastic behavior and allows a rapid interface deformation at lower pressures, in much the same manner as does recrystallization; besides, the recrystallized metals allow higher diffusion rates during the process. Additionally, introducing elements with high diffusivity into the systems at the interface increases the diffusion, whose purpose is to avoid melting at the bonding interface. Combining diffusion bonding and severe plastic deformation is often used for aerospace structures [21].

Some research on diffusion bonding in different materials and metal matrix composites has been carried out. Velmurugan et al. [22] conducted diffusion bonding of Ti-6Al-4V and duplex stainless steel at low temperature (650 °C–800 °C for 30 min); a solid solution at the interface was formed, but the maximum shear strength was achieved at 750 °C. Ghosh and Chatterjee [23] used higher temperatures for bonding titanium and austenitic stainless steel, where some intermetallic phases were formed. Mahendran et al. [24] studied the diffusion bonding process in dissimilar metals, namely, Mg-Cu; they found the formation of oxide films on the magnesium surfaces and the formation of brittle metallic interlayers and oxide inclusions in the bonding region, which were hard to remove. Kitazono et al. [25] studied the plastic deformation of the AZ91 magnesium alloy by accumulative diffusion bonding. First, metallic sheets were stacked and uniaxially hot pressed at 673 K under a constant load of 15 ton during 1 h for obtaining samples of 2 mm in height; then the samples were cut and pressed again. They found a crystal structure orientation (texturing) that allowed to achieve better tensile strength and ductility than those of the base material. Afghahi et al. [26] joined 7075 aluminum alloy and AZ31 magnesium alloy at temperatures of 430–450 °C under a pressure of 10–35 MPa for 1 h, under a vacuum of 13.3 MPa. They found that the diffusion of aluminum atoms into the magnesium alloy prevailed and that the movement of the interface happens toward the aluminum alloy. Furthermore, some reactions at the interface were evidenced, identifying the intermetallic compounds $Al_{12}Mg_{17}$, Al_3Mg_2, and $\alpha(Al)$ solid solution.

In the case of metal matrix composites, the diffusion bonding process has been studied by several researchers, being an aluminum matrix reinforced with silicon carbide the most used. Zhang et al. [27] reinforced pure aluminum with SiC particulate (Al/SiCp) by the vacuum brazing process. The results revealed a weak bonding quality either in the SiCp/Al or SiCp/SiCp interfaces and consequently a decreasing strength of the brazed joints with the increase of the SiC content. Muratoğlu et al. [28] synthesized composites of pure aluminum reinforced with SiC particulate, studying the effect of homogenization and aging treatments on bonding properties. They found that the application of the heat treatments before and after the diffusion bonding process decreased the SiC particulate accumulation and increased the Cu concentration at the interface.

The diffusion bonding in some materials like aluminum-based alloys and composites can be challenging to achieve, owing to the oxide film that forms at their surface, which is chemically stable and very tough. In order to accomplish a good diffusion bonding, some researchers have chosen to break up the brittle and continuous oxide layer by plastic deformation [22–26]. Thereby, the metal-to-metal contact is promoted because of local disruption of the oxide film on both surfaces. Another option for avoiding the oxide film problem in some metals is to use a rough surface finish, which may lead to higher bond strengths compared to polished surfaces. A rougher surface may allow the plastic deformation on the asperities, leading to a local disruption of the oxide layer.

4.4.4 Microstructural and Structural Analysis of Composites

Figure 4.5 shows optical microscopy images of the cross section of Mg/MWCNTs and Al/MWCNTs composites processed by the sandwich technique. Figure 4.5a reveals that the Mg/MWCNTs composite has grains homogeneously distributed, which was induced by the annealing and hot rolling processes performed after compaction. A good diffusion between magnesium sheets can be observed, and some grains can be even seen passing through the interface between magnesium sheets. It is evident the grain twinning due to the post-processes. This twinning acts as a mechanism for the plastic deformation during the mechanical stress. Concerning the Al/MWCNTs composites, Fig. 4.5b reveals the aluminum matrix with dark areas corresponding to the $FeAl_3$ phase, which was identified by XRD analysis; as will be shown below, this phase is in low proportions in the aluminum alloy.

Figure 4.7 shows field emission scanning electron microscopy (FE-SEM) images of both kinds of composites. The close-ups correspond to the interface between the metal layers, i.e., the reinforced zones or the MWCNTs-rich zones. In general, these images show a good MWCNT dispersion for both kinds of composites (Mg/MWCNTs and Al/MWCNTs). Besides, the formation of MWCNT clusters was not observed, which allows to have a good load transference between the metal matrix and MWCNTs; this in turn leads to providing good mechanical properties, as will

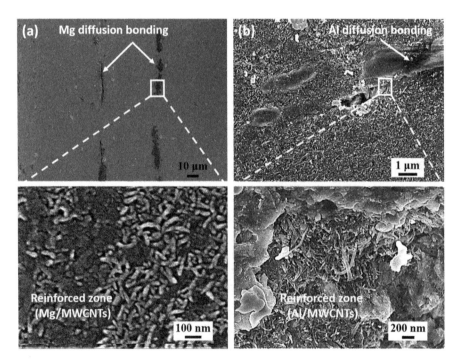

Fig. 4.7 FE-SEM images of (**a**) magnesium and (**b**) aluminum both reinforced with 0.5 wt% MWCNTs

4.4 Metal Matrix Composites Synthesis

be seen below. For the Mg/MWCNTs composite with 0.5 wt% MWCNTs (Fig. 4.7a), the reinforced zone shows some dark zones in the interface between magnesium sheets, which correspond to the formation of magnesium (MgO) and aluminum (Al_2O_3) oxides, as suggested by the energy-dispersive spectroscopy (EDS) mapping shown in Fig. 4.8. It is well known that magnesium corrodes via the oxidation in the presence of water or moisture, and then the formation of these oxides could be apparently due to the hydrogen and oxygen generated during the PVA pyrolysis. Regarding the Al/MWCNTs composites (Fig. 4.7b), the same behavior was found: the reinforced zone shows some dark zones in the interface between aluminum sheets, which correspond to the formation of aluminum oxide (Al_2O_3). Figure 4.9 presents the EDS mapping analysis of the interface for the Al/MWCNTs composite with 0.5 wt% MWCNTs. The presence of carbon, iron, and silicon was detected.

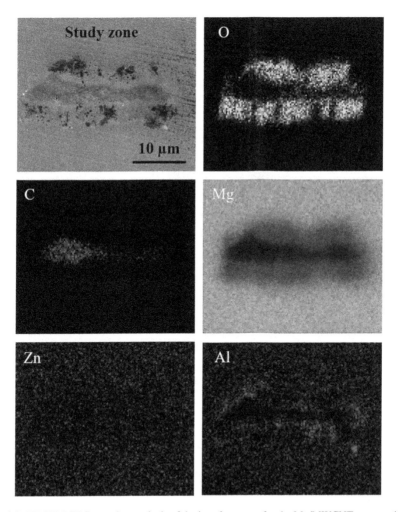

Fig. 4.8 FE-SEM-EDS mapping analysis of the interface zone for the Mg/MWCNTs composite

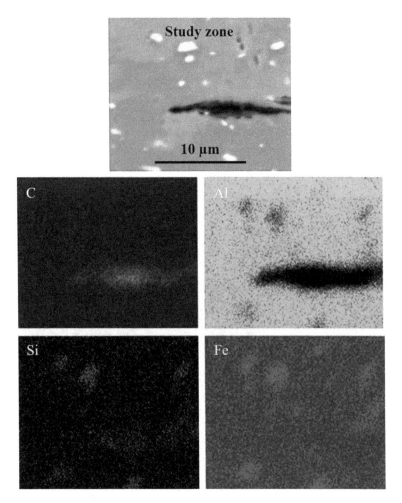

Fig. 4.9 FE-SEM-EDS mapping analysis of the interface zone for the Al/MWCNTs composite

Iron is an important element since it is the precursor of the AlFe$_3$ phase. It should be noted that the EDS analysis has an interaction volume greater than that shown in the images.

XRD results for both kinds of composites studied are shown in Fig. 4.10. The diffraction reflections in Fig. 4.10a clearly show the presence of the Mg and MWCNTs phases, suggesting the formation of the MgO and Mg$_{17}$Al$_{12}$ phases. There was no evidence of the Al$_2$O$_3$ and carbides formation between magnesium and MWCNTs during the diffusion bonding process; it is probable that these phases exist, but they could not be detected due to the equipment resolution. These results are in agreement with those found by the EDS analysis (Fig. 4.8). The diffractogram in Fig. 4.10b reveals the presence of iron carbide; the formation of the Fe-Al phase is characteristic of the 1100 aluminum alloy. Aluminum carbide was not detected by

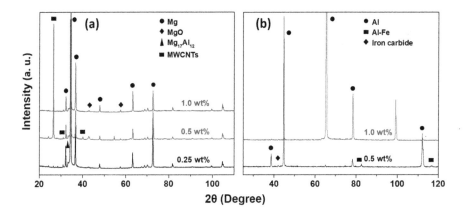

Fig. 4.10 XRD patterns of (**a**) magnesium and (**b**) aluminum reinforced with carbon nanotubes

XRD in these zones, which could indicate either a not significant interfacial reaction or resolution limitations of the instrument.

4.4.5 Dispersion Quantification of MWCNTs in Metal Matrix Composites Fabricated by the Sandwich Technique

In order to demonstrate the viability of the sandwich technique in the synthesis of light metal matrix composites, the dispersion quantification of MWCNTs was measured. A good dispersion of the reinforcement within the matrix brings about enhanced mechanical properties. The dispersion quantification was done using the statistical distribution model [29]. Cross-sectional and longitudinal sections were cut from samples and metallographically polished to be characterized by scanning electron microscopy. Representative images of 1 µm × 1 µm were divided into 10 × 10 grid lines along the horizontal and vertical directions. The spacing, both horizontal and vertical, between nearest MWCNTs was measured at each grid intersection using the Gwyddion free software [30]. In this method, the dispersion degree $D_{0.1}$ is deduced from the free-path (FP) distance distribution; a higher $D_{0.1}$ value indicates more spacing data close to the mean "u." A dispersion of 100% means that all of the entire reinforcement is equally spaced, i.e., different MWCNT contents can have the same value of the $D_{0.1}$ parameter. However, $D_{0.1}$ measures the amount of reinforcement that is at the same distance, regardless of its value. Certainly, any value of $D_{0.1}$, $D_{0.2}$, $D_{0.3}$, etc. can be chosen to describe the dispersion of the MWCNTs. The bigger the range, the bigger the value of D, i.e., there is a higher probability of finding more data in the range. As a result of this, low range values give information on the amount of MWCNTs that are equally spaced at a value near to the mean value (well dispersed); high range values give no information on the dispersion. Therefore, a range value of 0.1 was chosen in this work. It is worth noting that a value of 0.2 is also used by some authors [29].

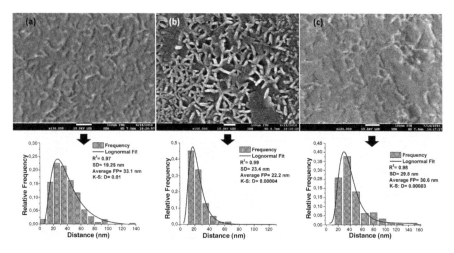

Fig. 4.11 FE-SEM images of the magnesium alloy reinforced with (**a**) 0.25, (**b**) 0.5, and (**c**) 1.0 wt% MWCNTs

Figure 4.11 shows FE-SEM images for the Mg/MWCNTs composites. The results of the dispersion quantification are plotted as histograms below the micrographs, which show a lognormal distribution. The dispersion degree was calculated from such histograms. The Kolmogorov-Smirnov (K-S) test was also done for comparing the cumulative distribution function with an equivalent distribution data. The test was done for each frequency distribution, and, for all cases, the statistics D was lower than the critical value calculated by tables with a significance value of 0.05. This means that the hypothesis for similar distribution functions is accepted. The micrographs in Fig. 4.11 correspond to the magnesium alloy reinforced with different content of MWCNTs. The carbon nanotubes are seen well dispersed into the matrix, with no formation of clusters. The calculated dispersion degree $D_{0.1}$ was 14.40%, 8.39%, and 9.01% for 0.25, 0.5, and 1.0 wt% MWCNTs, respectively. For methodology details refer to reference [8].

4.4.6 Microstructural Evolution Between Metallic Sheets in the Composites

A good understanding of the strengthening mechanism that happens at the interface between metal sheets and between MWCNTs and metal will lead to the comprehension of the mechanical behavior of the synthesized composites. Accordingly, in this section the interface analysis in magnesium and aluminum composites will be presented. Transversal sections of the reinforced

4.4 Metal Matrix Composites Synthesis

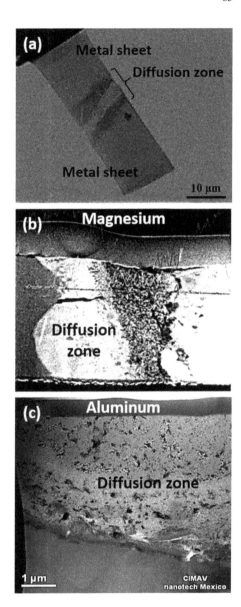

Fig. 4.12 (**a**) Lamella extracted close to the diffusion zone. Details from the diffusion zone of (**b**) Mg and (**c**) Al composites

zones of some samples were prepared by a focused ion beam (FIB) system. The analysis was focused on transmission electron microscopy (TEM) for both magnesium and aluminum composites. A detailed analysis will be discussed in Chap. 9.

Figure 4.12 presents TEM images of the diffusion zone for magnesium and aluminum composites, where the bonding between metal sheets is evidenced. Figure 4.13 shows bright field TEM images for magnesium composites reinforced

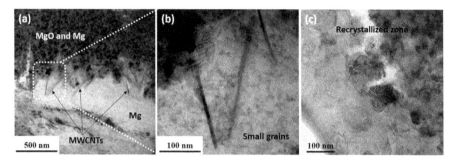

Fig. 4.13 Bright field TEM images for magnesium reinforced with 0.25 wt% MWCNTs. (**a**) transition zone, (**b**) detail of MWCNTs, and (**c**) detail of a recrystallized zone

with 0.25 wt% MWCNTs, which were taken at the transition zone, namely, where the diffusion zone ends. Figure 4.13a evidences the presence of MWCNTs, magnesium (bright phase) and magnesium oxide (dark phase). Figures 4.13a–c reveal that the MWCNTs are well immersed into the magnesium matrix. Additionally, the presence of MWCNTs, Mg, and MgO is evidenced; MgO is seen as small grains in the transition zone.

For aluminum reinforced with MWCNTs, the findings were similar. Figure 4.14 shows TEM images from the diffusion zone, where the MWCNTs can be seen well embedded into the aluminum matrix, with no evidence of cluster formation or MWCNTs damage. As is already known, these features are essential for a good load transfer and thereby to improve the mechanical properties of the metal matrix composites. These results clearly demonstrate that the sandwich technique is much less prone to induce reinforcement damage, since it does not use high mechanical energies compared with conventional composite manufacturing processes derived, for instance, from powder metallurgy.

4.4.7 Bulk Mechanical Properties of Metal Matrix/MWCNTs Composites

The mechanical properties of all synthesized composites were measured by tensile tests. Figure 4.15 shows the stress-strain curves and the effect of the MWCNTs content on the yield and ultimate stresses; these results correspond to the average of five tests for each composite. As can be observed, the addition of MWCNTs significantly improved the yield strength and the ultimate tensile strength, as well as the elastic modulus (qualitatively estimated). The increment in properties with the MWCNT addition definitely reflects a good interfacial bonding between MWCNTs and matrix. From these findings, it can be inferred that a good load transfer from the matrix to the MWCNTs was achieved, driven by a good dispersion and good interface between magnesium/aluminum and MWCNTs, as was discussed in previous

4.5 Conclusions

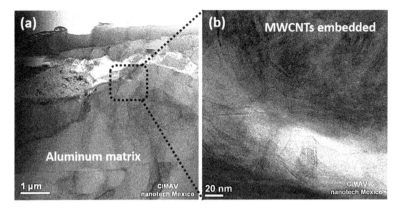

Fig. 4.14 (**a**) TEM image of aluminum reinforced with 0.5 wt% MWCNTs and (**b**) detail of (**a**)

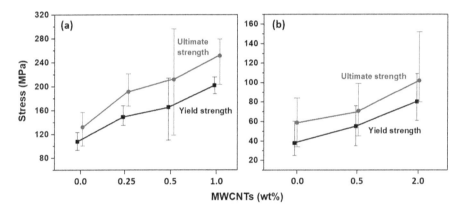

Fig. 4.15 Bulk mechanical properties for the composites fabricated by the sandwich technique (**a**) magnesium composites and (**b**) aluminum composites

sections. It is also mandatory to have a good interaction between the magnesium/aluminum layers and the "reinforced" layers, which in part is obtained during the diffusive process; thus, a gradual transition of properties, especially the elastic modulus, was obtained. In our synthesized composites, the sheet delamination for both kinds of composites was not evidenced.

4.5 Conclusions

Mg/MWCNTs and Al//MWCNTs composites were synthesized by a new process named "sandwich technique." This technique is particularly interesting because it produces a substantial increase in the mechanical properties due to the dispersion and in some cases the alignment of the MWCNTs into a metal matrix, without causing damage to the reinforcing material. Different strengthening mechanisms were identified, which corroborate the improvement in mechanical properties. The major parameters that affect the strengthening are MWCNT dispersion, the deformation during the processing, and the interface between MWCNTs and the metal matrix. Another behavior that allowed a strengthening of the composites synthesized by the sandwich technique was the grain size reduction induced by the processing. In addition, the presence of MWCNTs into the metal matrix led to the grain refinement.

Even if the sandwich technique has strong deformation processes (hot compaction and hot rolling) that could negatively impact the material properties, MWCNT damage was not identified, and dense composites were obtained. Thus, a good load transfer between the metal matrix and the MWCNTs was demonstrated. On the other hand, a higher degree of deformation is expected to increase the extent of alignment of MWCNTs along the hot rolling direction; however, the MWCNT alignment decreased during the hot compaction processing.

References

1. Salas, W., Alba-Baena, N., & Murr, L. (2007). Explosive shock-wave consolidation of aluminum powder/carbon nanotube aggregate mixtures: Optical and electron metallography. *Metallurgical and Materials Transactions A, 38*(12), 2928–2935.
2. Eizadjou, M., et al. (2008). Investigation of structure and mechanical properties of multilayered Al/Cu composite produced by accumulative roll bonding (ARB) process. *Composites Science and Technology, 68*(9), 2003–2009.
3. Garcia, E. J., Wardle, B. L., & Hart, A. J. (2008). Joining prepreg composite interfaces with aligned carbon nanotubes. *Composites Part A: Applied Science and Manufacturing, 39*(6), 1065–1070.
4. Yu, S., Tong, M. N., & Critchlow, G. (2010). Use of carbon nanotubes reinforced epoxy as adhesives to join aluminum plates. *Materials & Design, 31*, S126–S129.
5. Avedesian, M. M., & Baker H. (1999). *ASM specialty handbook: magnesium and magnesium alloys*. ASM international.
6. Campbell Jr, F. C. (2011). *Manufacturing technology for aerospace structural materials*. Elsevier.
7. Chen, Z., et al. (2018). Multilayered carbon nanotube yarn based optoacoustic transducer with high energy conversion efficiency for ultrasound application. *Nano Energy, 46*, 314–321.
8. Isaza, M. C. A., et al. (2018). Dispersion and alignment quantification of carbon nanotubes in a polyvinyl alcohol matrix. *Journal of Composite Materials, 52*(12), 1617–1626.
9. Noor, N., et al. (2018). Review on carbon nanotube based polymer composites and its applications. *Journal of Advanced Manufacturing Technology (JAMT), 12*(1), 311–326.
10. Vázquez-Moreno, J. M., et al. (2019). Preparation and mechanical properties of graphene/carbon fiber-reinforced hierarchical polymer composites. *Journal of Composites Science, 3*(1), 30.
11. Enqvist, E., et al. (2016). The effect of ball milling time and rotational speed on ultra high molecular weight polyethylene reinforced with multiwalled carbon nanotubes. *Polymer Composites, 37*(4), 1128–1136.
12. Guo, J., et al. (2018). A new finding for carbon nanotubes in polymer blends: Reduction of nanotube breakage during melt mixing. *Journal of Thermoplastic Composite Materials, 31*(1), 110–118.
13. Verma, P., et al. (2015). Excellent electromagnetic interference shielding and mechanical properties of high loading carbon-nanotubes/polymer composites designed using melt recirculation equipped twin-screw extruder. *Carbon, 89*, 308–317.

14. Feng, W., et al. (2003). Well-aligned polyaniline/carbon-nanotube composite films grown by in-situ aniline polymerization. *Carbon, 41*(8), 1551–1557.
15. Gao, J., He, Y., & Gong, X. (2018). Effect of electric field induced alignment and dispersion of functionalized carbon nanotubes on properties of natural rubber. *Results in Physics, 9*, 493–499.
16. Shi, Y.-D., et al. (2018). Low magnetic field-induced alignment of nickel particles in segregated poly (l-lactide)/poly (ε-caprolactone)/multi-walled carbon nanotube nanocomposites: Towards remarkable and tunable conductive anisotropy. *Chemical Engineering Journal, 347*, 472–482.
17. Huang, Z.-M., et al. (2003). A review on polymer nanofibers by electrospinning and their applications in nanocomposites. *Composites Science and Technology, 63*(15), 2223–2253.
18. Lu, K., et al. (1996). Mechanical damage of carbon nanotubes by ultrasound. *Carbon, 34*(6), 814–816.
19. Mukhopadhyay, K., Dwivedi, C. D., & Mathur, G. N. (2002). Conversion of carbon nanotubes to carbon nanofibers by sonication. *Carbon, 8*(40), 1373–1376.
20. Medina Escobar, S. A., Isaza Merino, C. A., & Meza Meza, J. M. (2015). Mechanical and thermal behavior of polyvinyl alcohol reinforced with aligned carbon nanotubes. *Matéria (Rio de Janeiro), 20*(3), 794–802.
21. Rusnaldy, R. (2001). Diffusion bonding: An advanced of material process. *Rotasi, 3*(1), 23–27.
22. Velmurugan, C., et al. (2016). Low temperature diffusion bonding of Ti-6Al-4V and duplex stainless steel. *Journal of Materials Processing Technology, 234*, 272–279.
23. Ghosh, M., & Chatterjee, S. (2002). Characterization of transition joints of commercially pure titanium to 304 stainless steel. *Materials Characterization, 48*(5), 393–399.
24. Mahendran, G., Balasubramanian, V., & Senthilvelan, T. (2009). Developing diffusion bonding windows for joining AZ31B magnesium and copper alloys. *The International Journal of Advanced Manufacturing Technology, 42*(7–8), 689–695.
25. Kitazono, K., Shimoda, Y., & Kato, S. (2013). Enhanced plastic deformation of magnesium alloy produced through accumulative diffusion bonding. *Materials Science Forum, 735*, 87–92.
26. Afghahi, S. S. S., et al. (2016). Diffusion bonding of Al 7075 and Mg AZ31 alloys: Process parameters, microstructural analysis and mechanical properties. *Transactions of Nonferrous Metals Society of China, 26*(7), 1843–1851.
27. Zhang, X., Quan, G., & Wei, W. (1999). Preliminary investigation on joining performance of SiCp-reinforced aluminium metal matrix composite (Al/SiCp–MMC) by vacuum brazing. *Composites Part A: Applied Science and Manufacturing, 30*(6), 823–827.
28. Muratoğlu, M., Yilmaz, O., & Aksoy, M. (2006). Investigation on diffusion bonding characteristics of aluminum metal matrix composites (Al/SiCp) with pure aluminum for different heat treatments. *Journal of Materials Processing Technology, 178*(1–3), 211–217.
29. Luo, Z., & Koo, J. H. (2007). Quantifying the dispersion of mixture microstructures. *Journal of Microscopy, 225*(2), 118–125.
30. Gwyddion: Free SPM data analysis software. (2019). Scanning Probe Image Processor. Available from: http://gwyddion.net/

Chapter 5
Severe Plastic Deformation

Abstract This chapter presents the production of light alloys (Ti-Mg) by severe plastic deformation. Alloys with different Mg content were synthesized by high-pressure torsion straining, technique that allows to obtain ultrafine-grained materials in both equilibrium and nonequilibrium, changing the chemical, thermodynamic, and mechanical properties of the materials. In the Ti-Mg system, the solid solubility of Mg in Ti is less than 2 at% and high-pressure torsion technique proved to be a useful way for the synthesis of Ti-Mg alloys supersaturated with Mg. Therefore, this severe plastic deformation technique is an effective route for the processing of light alloys and composites.

5.1 Introduction to Severe Plastic Deformation

The techniques of severe plastic deformation (SPD) are experimental processes of metal forming where high plastic deformations are applied on a bulk process [1]. Metals of ultrafine grain (100–300 nm) can be obtained with SPD, having significant effects on their functional and mechanical properties [2]. The aim of these techniques is to produce lightweight materials of high mechanical strength [1]. Some examples of these techniques are equal-channel angular pressing (ECAP), accumulative roll bonding (ARB) process, repetitive corrugation and straightening (RCS), multi-directional forging (MDF), twist extrusion (TE), cyclic extrusion and compression (CEC), severe torsion straining (STS), cyclic closed-die forging (CCDF), super short multi-pass rolling (SSMR), multi-pass high-pressure sliding (MP-HPS), and high-pressure torsion (HPT) [1–3].

5.2 High-Pressure Torsion

High-pressure torsion (HPT) is a processing technique introduced initially by Bridgman [4], going through the arrangement of Erbel's device for ring-shaped samples [5], that of Valiev et al. for thin disk-shaped samples [2], to that of Sakai et al. for bulk samples [6, 7]. For the HPT process, thin disk-shaped samples are required, which are subjected to a high confining pressure and then strained in

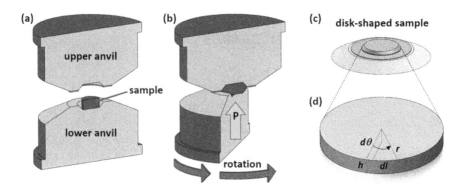

Fig. 5.1 Virtual models of the HPT process

torsion [8]. Experiments show that HPT is especially effective for producing materials with extremely small grain sizes [1, 2, 6–11].

The principle of the HPT technique involves the steps shown schematically in Fig. 5.1 [2, 10, 11]. The disk-shaped sample is placed between two anvils (Fig. 5.1a), where it is subjected to a compressive pressure (P) (Fig. 5.1b), either at room temperature or at high temperature. At the same time that the pressure is held constant, the sample is subjected to a torsional strain imposed through rotation of the lower anvil [11]. Due to the configuration of the process, a quasi-hydrostatic pressure is applied under compression, resulting in high stress values but preventing the sample from breaking (Fig. 5.1c). Considering a sample with radius r and thickness h (Fig. 5.1d), to which an infinitely small rotation, $d\theta$, and a displacement, $dl = rd\theta$, are applied, the incremental shear strain is calculated according to Eq. 5.1.

$$d\gamma = \frac{dl}{h} = \frac{rd\theta}{h} \qquad (5.1)$$

If the thickness of the sample is independent of the rotation angle, $\theta = 2\pi N$, the shear strain can be calculated by Eq. 5.2.

$$\gamma = \frac{2\pi Nr}{h} \qquad (5.2)$$

where N is the number of turns made to the sample [2, 12, 13]. Thus, during the HPT process, the torsional strain depends on the radius and thickness of the sample, as well as the number of turns applied.

The samples processed by this technique present submicron- or nano-grained microstructure and high mechanical strength [1]. As an example of the processing of light alloys and composites by severe plastic deformation, results of Ti-Mg alloys processed by HPT will be presented below.

5.3 Titanium-Magnesium Alloys

Nowadays, there is a necessity for better materials that meet the specifications of several industries. This need is being met with the development of new materials that are appropriated for both behavior and properties. A suitable study of different properties and behavior of materials leads to a useful scientific knowledge for an adequate selection of materials before they are put into service, which guarantees optimal operation and risk reduction. Despite the introduction of new materials in recent years, metals continue to be important in various structures owing to their mechanical strength, rigidity, hardness, and tolerance to high temperatures.

Titanium alloys are widely used in the aerospace industry, where low density, high strength-to-weight ratio, and good corrosion resistance are required. One way to reduce the weight of structural components made from Ti could be to add lighter alloying elements, such as Li, Ca, and Mg. However, these elements have a melting point well below that of Ti, which makes them virtually insoluble.

Magnesium alloys are some of the lightest metallic materials known, which compete with aluminum alloys for structural applications. However, magnesium alloys are not usually as strong as those of aluminum, having a lower modulus of elasticity. The biggest obstacle to the use of magnesium alloys is their extremely poor corrosion resistance.

Due to the above information, Ti-Mg alloys are perfect candidates for structural applications where high strength and low density are required. At present there are no commercial Ti-Mg alloys, much less those that contain high Mg percentage. The equilibrium phase diagram of the Ti-Mg system (Fig. 5.2) shows that the mutual solubility of magnesium and titanium is less than 2 at%. Furthermore, there are no reported intermetallic compounds. Due to the limited solubility of Mg in Ti and a difference in their melting points (Ti = 1668 °C, Mg = 650 °C), it is impossible to manufacture Ti-Mg alloys by conventional metallurgy (electric arc casting, rapid solidification, and solid-state processing) [14–16]. Nevertheless, it is expected that the combination of Ti and Mg by a suitable method can result in a significantly lighter alloy than other titanium-based alloys.

Some researchers have synthesized satisfactorily Ti-Mg alloys by using mechanical alloying as a fundamental basis [15, 17–20]; equilibrium and nonequilibrium phases can be obtained by this technique. Synthesized phases out of equilibrium include supersaturated solid solutions, metastable crystalline phases, quasicrystals, nanostructures, and amorphous alloys [21, 22]. The processing of systems out of equilibrium has drawn the attention of scientists and engineers due to the possibility of producing and improving materials, without considering conventional methods [22]. When using out-of-equilibrium system manufacturing techniques, the large difference in the melting points between titanium and magnesium is no longer considered a problem, and their solubility can be substantially increased [23]. Techniques such as direct current magnetron sputtering [24], mechanical alloy combined with mechanochemical processes [17], mechanical alloy combined with physical vapor deposition [20], spark plasma sintering [18], and high-pressure

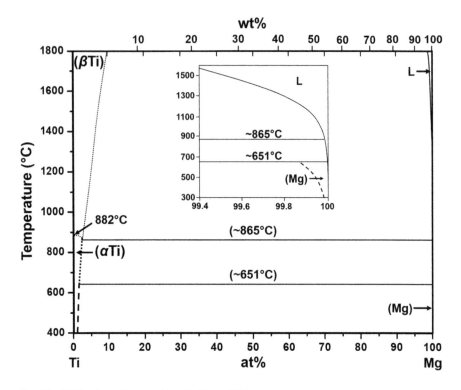

Fig. 5.2 Ti-Mg phase diagram. (Adapted from [25])

torsion [9] have extended the solid solubility achieving Ti-Mg alloys supersaturated with Mg. Under this context, in this chapter the production of Ti-Mg alloys by high-pressure torsion straining will be described.

5.4 Ti-Mg Alloys Synthesized by HPT

As an overview of what this study consisted of, Fig. 5.3 illustrates the process followed to obtain Ti-Mg alloys by high-pressure torsion straining, as well as for their characterization through different techniques. The characterization of alloys and composites typically involves a variety of complementary techniques, such as X-ray diffraction (XRD), optical microscopy, scanning electron microscopy (SEM), transmission electron microscopy (TEM), and energy-dispersive spectroscopy (EDS), among others. This combination is key for having a good understanding of the structure, microstructure, and properties of the alloyed products. A description of these techniques will be presented in Chap. 8.

For the example presented here, the raw materials were provided by High Purity Chemicals Ube Industries, with a purity of 99.9% for Ti and 99.5% for Mg, with a

5.4 Ti-Mg Alloys Synthesized by HPT 73

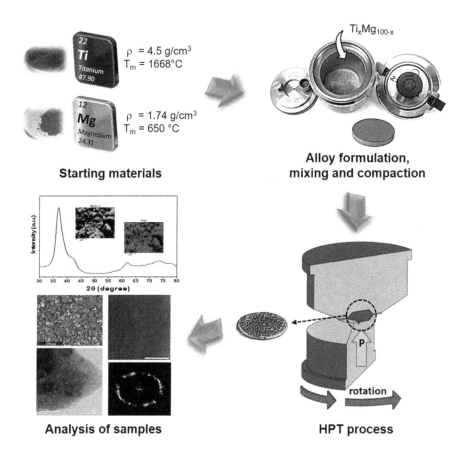

Fig. 5.3 Schematic production route of Ti-Mg alloys by HPT and their characterization

particle size of 40 µm and 180 µm, respectively. The powders were characterized by SEM using a Hitachi SU3500 microscope operated at 15 kV. Figure 5.4 presents secondary electron scanning electron microscopy (SE-SEM) images of both powders at different magnifications, where heterogeneous morphology and particle size can be observed.

The powders were also characterized by XRD using a Rigaku Ultima IV diffractometer with Cu-Kα (0.1542 nm) radiation at 40 kV and 40 mA. Figure 5.5 shows diffractograms for Ti and Mg, which have sharp and well-defined peaks, indicative of crystalline materials. The XRD patterns allowed to calculate the lattice parameters of both powders. After indexing the diffractograms, it was confirmed that both elements possess hexagonal close-packed (hcp) structure, according to JCPDS 00–044-1294 for Ti and 00–035-0821 for Mg, with a space group P63/mmc. As can be seen, the lattice parameters of Mg are greater than those of Ti, which implies a different crystalline cell volume: 35.9 Å^3 for Ti and 43.5 Å^3 for Mg. Figure 5.5 also

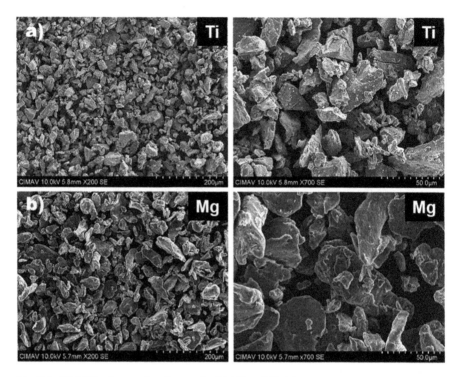

Fig. 5.4 SE-SEM micrographs of as-received (**a**) Ti and (**b**) Mg

shows schematic representations of the crystalline structure for both elements, generated by VESTA 3.4.0 [26] and CaRIne Crystallography 3.0 [27] software.

Figure 5.6 displays the polymorphic transformation of Ti. Note that, as stated in the pressure-temperature (P-T) phase diagram shown in Fig. 5.6a, Ti has an hcp structure at ambient temperature and pressure, which corresponds to the α phase. As the temperature increases, the α phase transforms to a body-centered cubic (bcc) structure at 882 °C [28], corresponding to the β phase. The α phase may also transform into a simple hexagonal structure (hex) at room temperature by increasing the pressure above 2 GPa, which corresponds to the ω phase. The XRD reflections of these phases can be seen in Fig. 5.6b, which were generated in the Powder Cell software [29]. Concerning Mg, it has an hcp structure and does not present allotropic transformations even by changing temperature or pressure.

The formulation of the alloys was carried out as follows: Ti and Mg powders were weighted inside a globe box filled with high-purity argon gas, in order to make nominal compositions Ti_xMg_{100-x} (x = 0, 25, 50, 75 and 100 at%). In some experiments, 4 wt% of stearic acid was added to the powders, to evaluate its effect on the HPT process. It is well known that the use of organic compounds can act as lubricant or surfactant to reduce the size of the particles, producing interdiffusion among them and contributing in this way to the formation of alloys [21]. A planetary P6 ball mill operated at 200 rpm was used to mix the powders for 2 h. The mixed

5.4 Ti-Mg Alloys Synthesized by HPT

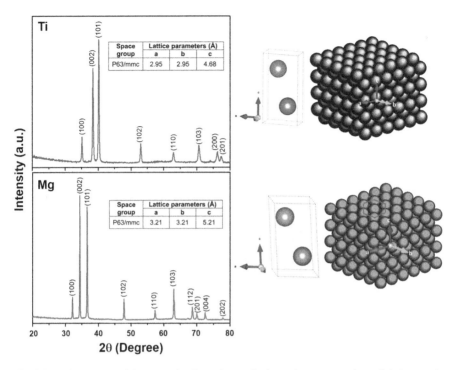

Fig. 5.5 XRD patterns of the as-received powders and schematic representations of their crystalline structure

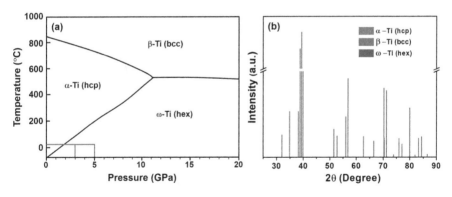

Fig. 5.6 (a) Schematic P-T phase diagram of titanium, adapted from [28]. (b) XRD reflections of the α, β, and ω phases of Ti obtained in Powder Cell software [29]

powders were cold consolidated using 40 MPa to obtain disks of 10 mm in diameter and 1 mm in thickness.

For the HPT processing, the disk-shaped samples were placed between two anvils. Then, a compressive pressure (P) of 3 or 5 GPa was applied, and simultaneously an intense shear strain was induced by rotating the lower anvil. The number of turns applied was $N = 10$ or 30 at a rotating speed of 0.2 rpm and $N = 100$ at 1 rpm. The experiments were performed in air atmosphere and at room temperature. The final material was a disk-shaped product with 10 mm in diameter and 0.85 mm in thickness. Samples were polished with SiC emery paper up to grit 4000 and then using 1-μm diamond paste to get a mirror finishing. Ethanol was used as a lubricant in all the metallographic preparations.

5.5 Characterization of Ti-Mg Alloys

The analysis of the synthesized samples was first done by X-ray diffraction (XRD) to investigate the formation of the alloys. The XRD method is very sensitive to composition, but additions around 1% are difficult to identify [30]. That is, if the samples had, for instance, any contaminant below such a value, it would not be detected by this technique.

Figure 5.7 presents the XRD results of the $Ti_{50}Mg_{50}$ composition before and after being processed by HPT. The XRD reflections of the as-blended powders correspond to those of elemental Ti and Mg. As identified in Fig. 5.5, both elements possess an hcp structure. When the powders were processed by HPT applying 5 GPa and 10 turns, the formation of the ω phase begins to become visible. According to the P-T phase diagram of Ti (Fig. 5.6a), a pressure of 5 GPa is enough to cause the transformation from the α phase to the ω phase. As the number of turns increases ($N = 30$), the intensities of the planes (110) and (201) corresponding to the ω phase

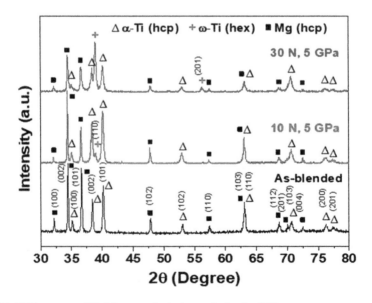

Fig. 5.7 XRD patterns of $Ti_{50}Mg_{50}$ samples before and after the HPT process

5.5 Characterization of Ti-Mg Alloys

of Ti increased significantly. However, a high concentration of α-Ti phase and Mg is still present. Taking into account that the aim is to form an alloy between Ti and Mg, the application of 5 GPa promoted the formation of the ω phase, but not that of the Ti-Mg alloy.

Due to the above, it was decided to reduce the HPT pressure at $P = 3$ GPa to study the effect of this parameter on the formation of the alloy. Besides, considering that the number of turns $N = 30$ was not enough to form the Ti-Mg alloy, it was increased to $N = 100$; it was expected that as the number of turns increased, the plastic deformation would increase to mechanically favor the formation of the alloy.

Similar to processes like mechanical alloying, the HPT process introduces a high density of defects to the material; hence the use of a process control agent (PCA) may be necessary to act as lubricant and produce finer particles by preventing cold welding. Thus, samples with and without stearic acid as a PCA were formulated and subjected to the HPT process.

Figure 5.8 shows the XRD patterns for the $Ti_{50}Mg_{50}$ samples processed under the aforementioned conditions. The diffractogram of the sample with no stearic acid is similar to those of Fig. 5.7, identifying the α-Ti (hcp) and Mg (hcp) phases. It could be thought that the energy and defects introduced by HPT were not enough to cause the phase transformation. In contrast, the diffractogram of the sample with stearic acid, having broad and short peaks, reveals the transformation from α-Ti (hcp) and Mg (hcp) phases to a nonequilibrium fcc phase. This finding indicates that the stearic acid acted as a process control agent to drive the phase transformation and achieve the alloy formation.

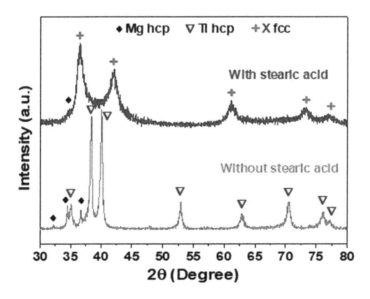

Fig. 5.8 XRD patterns of $Ti_{50}Mg_{50}$ samples

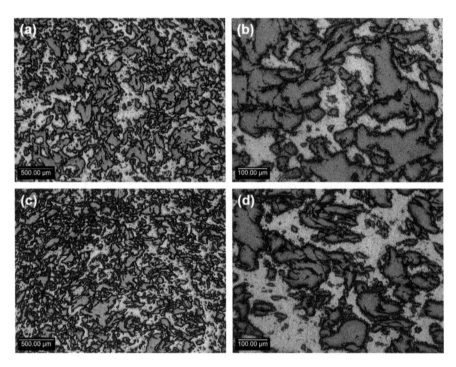

Fig. 5.9 Optical microscopy images of the samples processed by HPT as a function of the number of turns: (**a**) and (**b**) $N = 1$, (**c**) and (**d**) $N = 10$

The phase transformation from hcp to fcc is attributed to the heavy plastic deformation that occurs during the high-pressure torsion straining. When stearic acid was used during the HPT process, high refinement in the particle size and good interdiffusion of the elements were attained. The hcp and fcc structures are the closest packing structures, and the arrangements of atoms are similar. Therefore, the stacking of (111) planes in the sequence ABCABCABC… will result in the formation of an fcc structure. On the other hand, the stacking of (0001) planes in the sequence ABABAB… or ACACAC… will result in the formation of an hcp structure. Thus, the introduction of stacking faults at regular intervals produces the transformation hcp $\leftrightarrow$ fcc. This behavior was observed in other hcp materials like Zr and Co [19, 31].

An atomistic simulation study performed by Zheng et al. [32] to nanocrystalline hcp cobalt revealed that when the grain size is below 20 nm and a stress is applied, the density of (Shockley) partial dislocations increases as the strain increases, leaving an accumulation of stacking faults. Another study suggests that the Gibbs free energy for the fcc phase (in Ti and Zr) can be lower than that for the hcp phase when the grain size is in the range of a few nanometers [33].

Figures 5.9 and 5.10 present the morphology evolution of the samples processed by HPT at several numbers of turns; the micrographs were taken using an Olympus GX71 inverted metallurgical microscope. In a general way, two phases are

5.5 Characterization of Ti-Mg Alloys

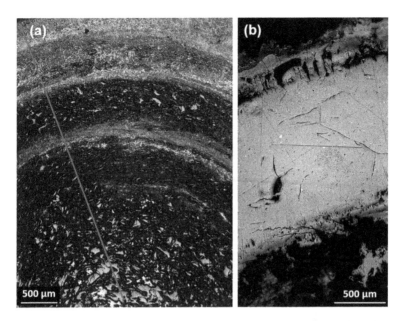

Fig. 5.10 Optical microscopy images of the samples processed by HPT as a function of the number of turns: (**a**) $N = 30$, (**b**) $N = 100$

identified, and the microstructure becomes more refined and homogeneous as the number of turns increases. The difference between samples processed with 1 turn (Figs. 5.9a–b) and 10 turns (Figs. 5.9c–d) is only the microstructure refinement. However, when the number of turns increases to $N = 30$ (Fig. 5.10a), differences in various zones of the sample can be seen. Indeed, the center of the sample has a poorly refined microstructure with coarse grains, similar to those in the sample with 1 turn. Zones in the half-radius of the sample have refined grains, being this refinement greater at the edge of the sample. This phenomenon is in agreement to that described by several authors [2, 7]. Finally, a more homogeneous and well-refined microstructure is obtained in the sample processed with 100 turns (Fig. 5.10b).

The results presented in Figs. 5.9 and 5.10 are a good example of the capabilities provided by optical microscopy. Nevertheless, for a better understanding of the microstructure characteristics, scanning electron microscopy (SEM) is recommended. Unlike a secondary electron (SE) signal that provides information on the morphology of samples (see images in Fig. 5.4), a backscattered electron (BSE) signal provides information on the chemical composition. This signal is sensitive to variations in the atomic number (Z) of the elements present on the interaction area. In a completely smooth surface, different grayscale will be seen depending on the amount of elements present. A lighter grayscale will indicate the presence of an element with a higher atomic number.

Figure 5.11 shows a general view of the samples using BSE-SEM, where it is possible to see differences in contrast indicating the presence of different elements or phases in the sample composition. These micrographs corroborate the results of

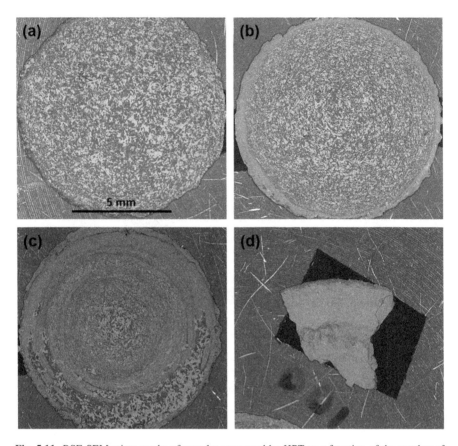

Fig. 5.11 BSE-SEM micrographs of samples processed by HPT as a function of the number of turns: (a) $N = 1$, (b) $N = 10$, (c) $N = 30$, (d) $N = 100$

optical microscopy (Figs. 5.9 and 5.10), observing also the presence of two phases: the light phase corresponds to Ti ($Z = 22$) and the dark phase to Mg ($Z = 12$). It is worth mentioning that the sample subjected to 100 turns was fractured during the metallographic preparation. It is well known that intense deformations cause the introduction of many residual stresses in the processed materials, which may originate their embrittlement; processes such as stress-relieving heat treatments help reduce this problem, before placing a component into service.

Observations by SEM at higher magnifications show some details of the samples. The morphology of the sample processed with 1 turn (Fig. 5.12) is homogeneous in the entire sample, from the center to the edge. The morphology of the sample processed with 10 turns (Fig. 5.13) is more refined than the previous one and homogeneous in all the samples except at its edge, where the microstructure refinement and heavy plastic deformation are noticeable. For the sample processed with 30 turns (Fig. 5.14), a huge difference in the microstructure refinement can be seen among the center, the half-radius, and the edge of the sample. Further, texturing in

5.5 Characterization of Ti-Mg Alloys

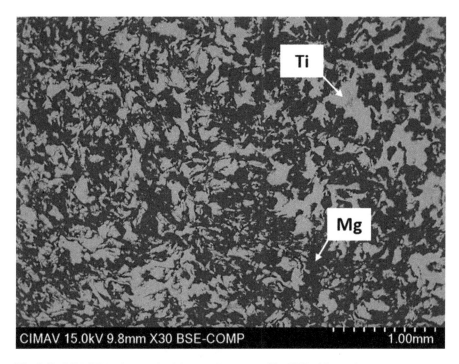

Fig. 5.12 BSE-SEM micrograph of the sample processed by HPT with $N = 1$

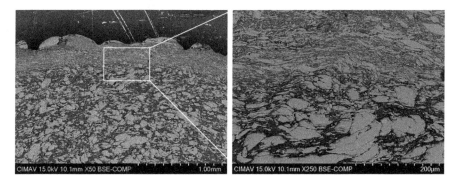

Fig. 5.13 BSE-SEM micrographs of the sample processed by HPT with $N = 10$

the direction of deformation of the anvils can be appreciated as one advances from the center to the edge. The sample processed with 100 turns (Fig. 5.15) has a very homogeneous and well-refined microstructure in the whole sample, and no differences in composition are identified.

In addition to morphological (SE-SEM) and compositional (BSE-SEM) observations, a chemical characterization of samples is possible to make with the use of energy-dispersive spectroscopy (EDS) coupled to scanning electron microscopy [34, 35]. The EDS technique detects X-rays emitted from the sample during the

Fig. 5.14 BSE-SEM micrograph of the sample processed by HPT with $N = 30$

beam collision with its surface to determine the elemental composition of the analyzed volume. The screened X-ray energy is characteristic of the element from which it was emitted due to the fact that each element possesses a unique atomic structure.

The spatial distribution of the elements in the samples processed by HPT was examined through SEM-EDS mappings. Figure 5.16 shows that the elemental composition of all samples corresponds to Ti, Mg, and O. The oxygen is difficult to avoid, since both Ti and Mg are avid for this element, which can be captured from the production of the elemental powders used as raw material and from the environment during the preparation, synthesis, and analysis of samples. A heterogeneous distribution of the constituents is observed in the sample processed with 1 turn (Fig. 5.16a). This distribution becomes more homogeneous as the number of turns increases to 10 and 30 (Figs. 5.16b–c), until reaching a total homogeneity, along with a well-refined microstructure, for the sample processed with 100 turns (Fig. 5.16d).

As a complementary analysis, Fig. 5.17 presents results of EDS elemental mappings, which allow to see how the elements are distributed along the matrix of samples. In Figs. 5.17a–b, it can be observed that the black areas of Ti match well with the existence of Mg, and the oxygen is homogeneously distributed along the matrix.

5.5 Characterization of Ti-Mg Alloys

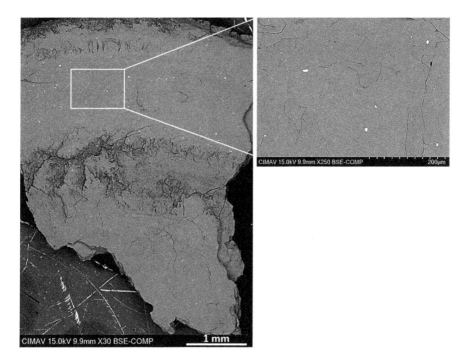

Fig. 5.15 BSE-SEM micrograph of the sample processed by HPT with $N = 100$

The presence of oxygen is more evident in Fig. 5.17c. The region enclosed by a white line indicates where the Ti is concentrated and where the Mg is absent. Then, because the oxygen is present in the same region, the formation of some titanium oxide is suggested; the rest of the analyzed area seems to be homogeneous in oxygen, suggesting the probable formation of some Ti and/or Mg oxide. This is to be expected because of the avidity of Ti and Mg for oxygen. Furthermore, during the HPT process, the oxygen crust of the powders is broken as the number of rotations increases, leaving the powders exposed to the oxygen of the atmosphere. However, this oxygen contamination and the probable formation of Ti_xO_y- and/or MgO-type oxides were not detected by XRD (Figs. 5.7 and 5.8). This is an indicative that, in case these oxides have formed, their percentage should be less than 1%, which is the detection limit of the XRD technique [30]. A good distribution of the elements and a well-refined microstructure can be confirmed in Fig. 5.17d, where the oxygen contamination seems to be null. The fact could be linked to the formation of the Ti-Mg alloy, which probably is not avid for O as the elements are.

Analysis by optical and scanning electron microscopies confirmed a well dissolution of Ti and Mg and microstructural refinement, while the analysis by X-ray diffraction proved the formation of the Ti-Mg alloy accompanied by a phase transformation. Transmission electron microscopy (TEM) is a powerful and sophisticated technique that allows the study of matter from a few nanometers. Furthermore, chemical analysis and mapping by EDS and selected area electron diffraction

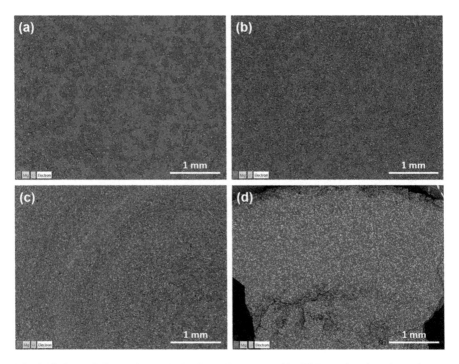

Fig. 5.16 SEM-EDS mapping analysis of samples processed by HPT as a function of the number of turns: (**a**) $N = 1$, (**b**) $N = 10$, (**c**) $N = 30$, (**d**) $N = 100$

(SAED) patterns can also be performed. A Hitachi HT 7700 microscope operated at 100 kV was used to characterize the sample processed with 100 turns. For this purpose, the sample was prepared by a focused ion beam (FIB) system (JEOL JEM 9320FIB), in order to obtain a thickness thin enough (~100 nm) to be properly analyzed by TEM. Bright and dark field images in Fig. 5.18 show the presence of a nanocrystalline material with a grain size between 7 nm and 17 nm. The nanograins are well distributed along the sample. The SAED pattern reveals characteristic rings of a polycrystalline material, whose indexing matches with an fcc structure, as previously detected by XRD. The SAED pattern was indexed by the Process Diffraction software [36].

Energy-dispersive spectroscopy (EDS) coupled to the transmission electron microscope was used to identify the chemical elements in the sample, in addition to their distribution (Fig. 5.19). Even at this scale, the results allow to see a homogeneous distribution of Ti and Mg along the analyzed area. These findings confirm that HPT, under specific conditions, is a suitable method to process light materials such as Ti-Mg alloys.

5.5 Characterization of Ti-Mg Alloys

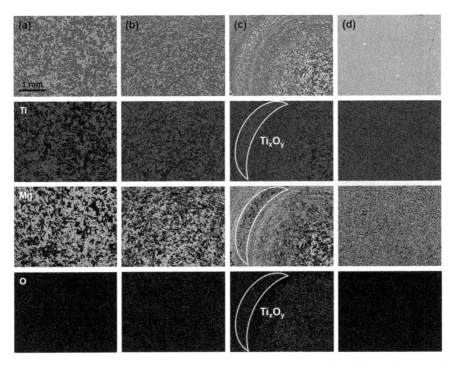

Fig. 5.17 SEM-EDS elemental mapping analysis of samples processed by HPT as a function of the number of turns: (**a**) $N = 1$, (**b**) $N = 10$, (**c**) $N = 30$, (**d**) $N = 100$

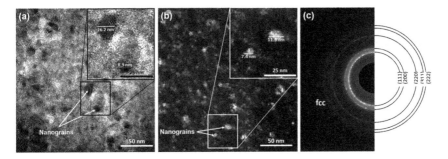

Fig. 5.18 TEM analysis of the sample processed by HPT with $N = 100$: (**a**) bright field, (**b**) dark field, and (**c**) indexed SAED pattern

In a general way, the TEM results led to confirm the phase transformation mechanism discussed above. The HPT process allowed the interdiffusion of the elements obtaining a good distribution of nanograins, with a grain size below 20 nm.

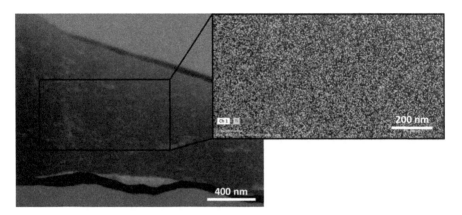

Fig. 5.19 TEM-EDS elemental mapping analysis of the sample processed by HPT with $N = 100$

5.6 Conclusions

The effectiveness of severe plastic deformation by the HPT technique to obtain light alloys (Ti-Mg) with an ultrafine- and nano-grained microstructure was demonstrated. The limited equilibrium solubility of Mg in Ti was greatly increased by HPT due to the heavy plastic deformation during their processing. As a consequence of this deformation, which was applied under controlled conditions of pressure and number of turns, a nonequilibrium fcc phase was stabilized by the introduction of stacking faults, which allowed the formation of Ti-Mg alloys.

References

1. Azushima, A., et al. (2008). Severe plastic deformation (SPD) processes for metals. *CIRP Annals, 57*(2), 716–735.
2. Valiev, R. Z., Zhilyaev, A. P., & Langdon, T. G. (2014). *Bulk nanostructured materials: Fundamentals and applications*. New Jersey: Wiley.
3. Tang, Y., et al. (2019). Multi-pass high-pressure sliding (MP-HPS) for grain refinement and superplasticity in metallic round rods. *Materials Science and Engineering A, 748*, 108–118.
4. Bridgman, P. (1935). Effects of high shearing stress combined with high hydrostatic pressure. *Physical Review, 48*(10), 825.
5. Erbel, S. (1979). Mechanical properties and structure of extremely strain hardened copper. *Metals Technology, 6*(1), 482–486.
6. Sakai, G., et al. (2005). Developing high-pressure torsion for use with bulk samples. *Materials Science and Engineering A, 406*(1–2), 268–273.
7. Sakai, G., et al. (2006). Application of high pressure torsion to bulk samples. *Materials Science Forum, 503*, 391–398.
8. Cantor, B. (2004). *Novel nanocrystalline alloys and magnetic nanomaterials*. Boca Raton: CRC Press.
9. Edalati, K., et al. (2015). Formation of metastable phases in magnesium–titanium system by high-pressure torsion and their hydrogen storage performance. *Acta Materialia, 99*, 150–156.

References

10. Zhilyaev, A., et al. (2003). Experimental parameters influencing grain refinement and microstructural evolution during high-pressure torsion. *Acta Materialia, 51*(3), 753–765.
11. Zhilyaev, A. P., & Langdon, T. G. (2008). Using high-pressure torsion for metal processing: Fundamentals and applications. *Progress in Materials Science, 53*(6), 893–979.
12. Xie, Z. L., et al. (2011). Microstructure and microhardness of copper subjected to high pressure torsion. In *Advanced Materials Research*. Trans Tech Publ.
13. Faraji, G., Kim, H. S., & Kashi, H. T. (2018). *Severe plastic deformation: methods, processing and properties*. San Diego: Elsevier.
14. Zheng, J. G., et al. (1997). Microstructure of vapour quenched Ti–29 wt% Mg alloy solid solution. *Journal of Materials Science, 32*(12), 3089–3099.
15. Liang, G., & Schulz, R. (2003). Synthesis of Mg-Ti alloy by mechanical alloying. *Journal of Materials Science, 38*(6), 1179–1184.
16. Haruna, T., et al. (2013). Corrosion resistance of titanium–magnesium alloy in weak acid solution containing fluoride ions. *Materials Transactions, 54*(2), 143–148.
17. Dargusch, M. S., & Keay, S. (2009). *Light metals technology 2009*. Trans Tech Publications Ltd.
18. Liu, Y., et al. (2015). Powder metallurgical low-modulus Ti–Mg alloys for biomedical applications. *Materials Science and Engineering: C, 56*, 241–250.
19. Suryanarayana, C., & Froes, F. (1990). Nanocrystalline titanium-magnesium alloys through mechanical alloying. *Journal of Materials Research, 5*(9), 1880–1886.
20. Hoffmann, I. (2014). Magnesium-titanium alloys for biomedical applications. In *Chemical and materials engineering*. Lexington: University of Kentucky.
21. Suryanarayana, C. (2001). Mechanical alloying and milling. *Progress in Materials Science, 46*(1–2), 1–184.
22. Suryanarayana, C., Ivanov, E., & Boldyrev, V. (2001). The science and technology of mechanical alloying. *Materials Science and Engineering A, 304*, 151–158.
23. Xu, Z., Song, G.-L., & Haddad, D. (2011). Corrosion performance of Mg-Ti alloys synthesized by magnetron sputtering. In *Magnesium technology 2011*. Springer.
24. Hieda, J., et al. (2015). In vitro biocompatibility of Ti–Mg alloys fabricated by direct current magnetron sputtering. *Materials Science and Engineering: C, 54*, 1–7.
25. Murray, J. (1986). The Mg– Ti (magnesium-titanium) system. *Bulletin of Alloy Phase Diagrams, 7*(3), 245–248.
26. Momma, K., & Izumi, F. (2011). VESTA 3 for three-dimensional visualization of crystal, volumetric and morphology data. *Journal of Applied Crystallography, 44*(6), 1272–1276.
27. Boudias, C., & Monceau, D. (1998). *CaRIne crystallography*. Compiègne: DIVERGENT SA.
28. Mei, Z.-G., et al. (2009). Density-functional study of the thermodynamic properties and the pressure–temperature phase diagram of Ti. *Physical Review B, 80*(10), 104116.
29. Kraus, W., & Nolze, G. (1996). POWDER CELL–a program for the representation and manipulation of crystal structures and calculation of the resulting X-ray powder patterns. *Journal of Applied Crystallography, 29*(3), 301–303.
30. Guinebretière, R. (2013). *X-ray diffraction by polycrystalline materials*. Somerset: Wiley.
31. Manna, I., et al. (2002). Formation of face-centered-cubic zirconium by mechanical attrition. *Applied Physics Letters, 81*(22), 4136–4138.
32. Zheng, G. P., Wang, Y. M., & Li, M. (2005). Atomistic simulation studies on deformation mechanism of nanocrystalline cobalt. *Acta Materialia, 53*(14), 3893–3901.
33. Xiong, S., et al. (2010). Size and shape dependent Gibbs free energy and phase stability of titanium and zirconium nanoparticles. *Materials Chemistry and Physics, 120*(2), 446–451.
34. Girão, A.V., Caputo, G., & Ferro, M.C. (2017). Application of Scanning Electron Microscopy–Energy Dispersive X-Ray Spectroscopy (SEM-EDS). In *Comprehensive analytical chemistry*. Elsevier (pp. 153–168).
35. Abd Mutalib, M., et al. (2017). Scanning Electron Microscopy (SEM) and Energy-Dispersive X-Ray (EDX) Spectroscopy. In N. Hilal et al. (Eds.), *Membrane characterization*. Amsterdam: Elsevier.
36. Lábár, J. L. (2005). Consistent indexing of a (set of) single crystal SAED pattern (s) with the ProcessDiffraction program. *Ultramicroscopy, 103*(3), 237–249.

Chapter 6
Additive Manufacturing

Abstract This chapter describes cutting-edge technologies that are being used in the manufacture of components from a complex three-dimensional design, as well as those technologies based on the remanufacturing and repair of metal parts for maximizing their work life. Among the technologies used by additive manufacturing, the sintering of metal powders for the construction of complex architectures has revolutionized various industries such as aerospace and automotive, where it is possible to design systems as complex as the imagination of designers allows, or feasibility in the manufacture of cooling systems with internal channels specifically adapted to wrap the geometry of components whose cooling requires maximum precision. The use of additive manufacturing not only extends to the manufacture of parts manufactured from a design produced in a computer system, but its versatility includes remanufacturing operations of commercial metal parts where it is possible to extend their functionality and operability for longer.

6.1 Introduction to Additive Metal Manufacturing

In the 1980s, researchers, from the University of Texas at Austin, produced the first component from the union of multiple layers of material [1]. This was the first step toward the development of more complex components and the beginning of the technology we know today as additive manufacturing (AM) [2]. Since then, this technology has been used as a tool for the product development based on a rapid prototyping process. This technology has been extended to the design and construction of parts for multiple applications and disciplines, allowing in their production geometries never before considered by conventional manufacturing techniques. Nowadays, additive manufacturing is being used by both the scientific and industrial sectors. The technique is a widely used process in the production of plastic components. However, these components find a limit in their applications when they are considered in the design and construction of structures or in applications where temperature represents a crucial parameter. For these reasons, the field of metallic materials is where this technology enhances the capabilities in the development of materials with interesting mechanical properties for the use at room temperature under severe workloads or in applications at temperatures beyond the point of degradation represented by plastic materials. Additionally, metals offer a wide range of interesting properties, such as mechanical strength and durability, and the development of AM-based technologies such as selective laser melting (SLM) and direct metal laser sintering (DMLS) allows their manipulation to design parts with an intricate design (Fig. 6.1).

Similar to its application in plastic products, the additive manufacturing in the production of metal components is carried out through different processes, which are generally based on the use of metal powders, wires, or sheets. Each of these processes is based on the addition of metals, in successive layers that allow obtaining a three-dimensional object from a digital file.

The potential of additive manufacturing has been rapidly disseminated, reaching the interest of scientific, technological, and governmental sectors around the world. The diffusion of its capabilities has been reflected in an increasing number of scientific articles year after year and has been considered by both the scientific and industrial sectors as an emerging technology, with wide possibilities in the field of the aerospace, automotive, biomedical, electronics, and instrumentation industries.

Although additive manufacturing is currently a technology away from serial production processes, its versatility makes it a valuable tool in prototype production, due to the freedom of design and the fine features it offers in the manufacture of channels or internal networks [3]. This feature gives additive manufacturing a competitive advantage over other manufacturing processes used in the production of various components. While currently AM does not offer the surface finish quality obtained through subtractive manufacturing, surfaces with a good finish are obtained compared to the processes involved in metal casting. Even though this feature is a disadvantage, hybrid systems are currently available and jointly involve metal addition systems for the AM production of specific objects, as well as tools for surface finishes based on the subtraction of material from the same object.

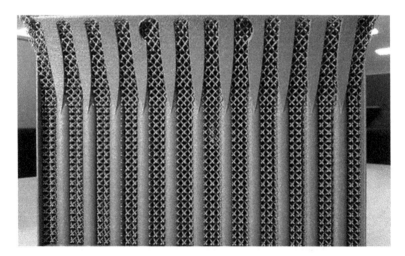

Fig. 6.1 3D metal sintered part produced by DMLS (*Photograph courtesy: Renishaw México, S. de R.L. de C.V*)

Unlike casting processes, the amount of materials available in AM processes is more limited. However, alloys available for medical and aerospace applications in the construction of unique parts are being used today [4–6]. Additionally, the unique characteristics of AM, in which powder metallurgy is used as the basis, provide a great opportunity for the synthesis of materials through the use of micro- and nanoparticles. These particles are being studied through several of the technologies associated with the AM, in which the spheroidism and particle size allow their fluidity through a nozzle. In addition, AM technologies such as cold spray (CS) and SLM involve studies about the dispersion of ceramic particles [7, 8].

The composite materials obtained through the technologies mentioned above are of special interest in the development of the automotive and aeronautical industries, among others. In this regard, this chapter focuses on the use of AM for the manufacture of light alloys and the emerging applications in the manufacture of parts from composite materials of metal matrix.

6.2 Characteristics and Advantages of Additive Metal Manufacturing

Additive manufacturing is capable of producing parts with complex shapes with acceptable mechanical properties for the development of materials for the automotive and aerospace industries, among others. The manipulation of metals in the powder form gives AM ample advantages in the production of composite materials that are currently in development and where the imagination of designers is the limit in the architectures that can be built (Fig. 6.2).

Although there is a high availability of technologies associated with AM, the most practical way of highlighting its characteristics is through the description of

Fig. 6.2 Schematic representation of a woven ball metal structure

Fig. 6.3 Multiple 3D metal sintered parts produced by DMLS (*Images courtesy: Renishaw México, S. de R.L. de C.V*)

powder bed-based processes. In these processes, a high-energy laser is used for sintering metal powder particles in successive layers to create objects in three dimensions. The formation of metal parts from powder layers is controlled digitally from 3D-created CAD designs. From a CAD system, a first and thin layer of powder is deposited where subsequently a laser sinters or melts the selected areas to a high degree of precision. This process is constantly repeated until the entire piece is formed. The pieces obtained through these technologies offer several advantages compared to subtractive manufacturing [9]:

1. Production of Multiple Pieces Simultaneously Unlike subtractive systems, AM systems enables the manufacture of multiple objects by including several laser systems as part of their architecture. In addition, AM is capable of printing components that are part of an assembly through a single printing process (Fig. 6.3).

2. Cost Reduction During Processing 3D printing is based on the use of a single tool or laser system that builds complex architectures of varying degrees of difficulty. The versatility of these technologies based on the sintering or fusion of multilayers avoids the use of additional tools in the manufacture of an object, regardless of the complexity of the design.

3. Design Freedom The AM allows the production of complex geometries that are considered expensive in their production through the use of conventional technologies based on subtractive manufacturing. The lack of this type of restrictions in the AM allows the designer to consider geometries with internal cuts or channels (Fig. 6.4) or structures based on three-dimensional internal networks, which allow to obtain an object with a considerable reduction in weight. This freedom in the design of pieces and in the control of their topography [10] also allows the manufacture of unique pieces for specific applications in which a cellular structure is possible. Furthermore, additional parts can be manufactured with the same degree of precision by basing their design on the specifications stored in a CAD file, in which it is also possible to add additional details regarding future changes or adaptations of the original design. Being based on a CAD design, these same changes can be sent from different parts around the world, to the production center for the creation of printed products.

Fig. 6.4 Multiple 3D metal sintered parts produced by DMLS (*Image courtesy: Renishaw México, S. de R.L. de C.V*)

4. Use of Multiple Materials From the freedom of design that AM allows, new forms with a complex construction can be produced. However, this freedom in design extends to the use of multiple materials during the manufacture of objects. Multiple materials can also be used in the manufacture of objects from different power supplies, where the end result can be a fully functional component composed of several segments of various materials [11].

5. Zero Waste The objects produced by AM are constructed from the successive deposit of the material, considering only the layers needed to form them . Although in many operations the component produced does not require additional manufacturing processes, sometimes and depending on the application of the object, the subsequent use of additional machining related to surface finishes is carried out. In the latter case, the amount of wasted material is minimal compared to that of conventional subtractive manufacturing [12].

6. Positive Impact on the Environment The complexity of the structures created by AM allows the production of lighter components for various industries. In these components, where it is possible to consider a cellular structure, a lightweight design focused on saving energy consumption is allowed. Also, some technologies allow the recycling of materials where the production of powders is possible through mechanical milling processes [13].

7. Scaling Through Its Use in Hybrid Systems Hybrid systems represent a solution to the problem of dynamic and continuous integration over time of different systems. Based on this concept, composite systems are currently available in the market, which allow for a consecutive and dynamic operation of additive manufacturing for the creation of objects, followed by subtractive manufacturing operations related to their final finishing. The combination of these technologies offers new and powerful capabilities to the industrial sector in relation to new manufacturing methods [14].

8. More Than Repair Is About Remanufacturing The most outstanding advantages of the AM are in the manufacture of objects in three dimensions from the successive union of metal layers. Moreover, the repair of tools is an area of innovation in which the AM shows strong capabilities when considering the reconstruction and obtaining of a product with a long operating life, at a lower cost and consequently avoiding its replacement [15].

6.3 Technologies Used in the AM of Light Alloys and Composites

In addition to metal powders, there are other power supplies that are used in AM technologies. In current systems, the power supply or the way it is processed must be taken into account. The latter is based on wires and sheets that allow the construction of complex architectures. In a complementary way, the feeding systems are based on a direct source or from a bed of powders, whose application is widely used in the remanufacturing of metal components.

6.3.1 Direct Metal Laser Sintering

Among the technologies that allow the construction of complex 3D architectures, the direct metal laser sintering (DMLS) is the most widely used technology in constant research and exploitation by the academic and industrial sectors. This is because this technology allows building parts with shapes whose degree of complexity is so high that it is only limited by the designer's imagination (Fig. 6.5). The DMLS technology is based on a feeding system whose base is a bed of dusts. Unlike a powder feed system, where the particles have the purpose of melting by using a higher-energy concentration in their union, sintering provides only the energy necessary to cause atomic diffusion among the particles that are in contact with each other, forming metallurgical bonds and providing mechanical stability to the whole.

Fig. 6.5 Automotive escape system configuration fabricated by DMLS (*Image courtesy: Renishaw México, S. de R.L. de C.V*)

6.3.2 Process and Equipment Used in DMLS

There is a wide variety of commercial elements and alloys available for the manufacture of different architectures from a design generated on a computer. During the DMLS process, a laser is used to sinter the first layer of metallic powder placed on the base plate, in order to generate the support structure (Fig. 6.6). Then the base plate moves down allowing a new layer of metallic powder to be deposited (Fig. 6.7). The sintering process of the subsequent layer follows the pattern generated by the computer design, successively until the manufacturing process is completed. Once the process is finished, the loose powder is removed.

6.3.3 Powders Used in DMLS

DMLS is a technology that relies on the use of metal powders for their later projection at a given pressure and speed. Most of the powders used in additive manufacturing are produced by atomization [16] (Fig. 6.8). The water atomization method is the most widely used for the production of metallic powders due to its low cost. However, the quality and size of the particles produced, related to their heterogeneity, are inferior in relation to other processes of atomization based on the use of gas, in addition to that it is not suitable for reactive materials such as titanium. The degree of reactivity, an improved flow in the particles, the increase in bulk density, as well as the homogeneity in the particle size are solved by using an atomization method where gases such as nitrogen or argon are used or by using plasma.

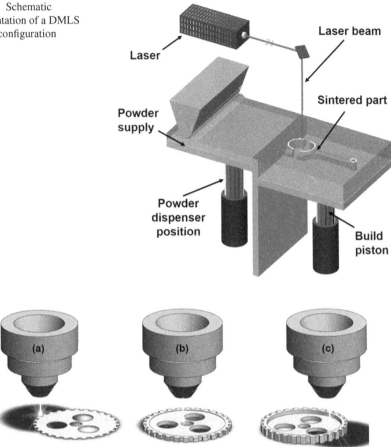

Fig. 6.6 Schematic representation of a DMLS system configuration

Fig. 6.7 Schematic representation of a gear produced by a DMLS system. (**a–c**) The component is formed layer by layer

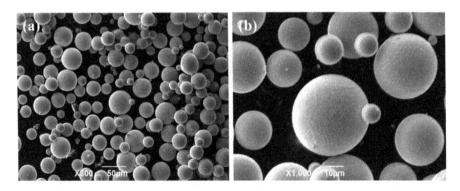

Fig. 6.8 Secondary electron SEM micrographs showing the morphology of metal powder particles produced via atomization as the feedstock material used in AM technologies (*Micrographs courtesy: Cummins Inc*)

6.3.4 Use of DMLS in the Plastic Injection Mold Industry

Even though the injection molds are mostly manufactured from steels, the incorporation of hybrid molds and molds made from light alloys is being used today. As light alloys in the manufacture of injection molds for the production of plastic parts, aluminum alloys are used, specifically those of aerospace grade. This is due to the fact that traditionally in those injection molds made from steel, the time required to cool a mold constitutes about 70% of the duty cycle for each part produced [3].

To further increase the productivity, the incorporation of conformed cooling channels is an alternative to improve the cooling time and the quality of the plastic part (Fig. 6.9). Traditional subtractive manufacturing methods are used in the design and manufacture of cooling channels, while channels manufactured by means of AM provide more precise temperature control of the molding cavity during the injection cycle.

The additive manufacturing and handling of metal powders allow the manufacture of parts of high interest in the tools industry [17]. This is because it allows considering architectures for new applications and new materials with interesting physical characteristics, which represent a competitive advantage in terms of saving machining time and increased performance in heat dissipation. This technology progresses rapidly in the injection mold industry, where component fabrications with specific applications are currently considered by the scientific and business sector [18].

In the conformal cooling, a preliminary analysis is carried out to determine the conditions that maximize the cooling of the plastic parts produced. In its implementation using AM techniques, an extensive analysis by computational fluid dynamics (CFD) is additionally required. It is possible to obtain an optimized cooling system by performing computer simulations of the flow inside the cavity and a cooling efficiency analysis of the mold and, in particular, of the part. Consequently, the plastic part will cool in a more homogeneous trend, which can shorten what is often the longest part of the mold cycle time setting. In addition, this can simultaneously impact the reduction of residual stresses and improve the quality of the piece, especially in the common problem areas of distortion and deformation.

Although conformal cooling applied to the manufacture of injection molds is not a new technology, the design, optimization, and use of new materials for manufacturing conforming cooling channels through the various AM methodologies are a very challenging research problem. Therefore, a multidisciplinary research team is needed together with engineers, designers, operators, and personnel involved in the manufacture of molds. Due to its practical uses, the design of the conforming cooling channels must meet the high standards of the global injection molding industry.

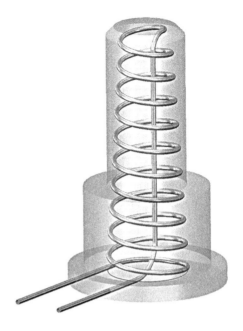

Fig. 6.9 Schematic representation of a conformal cooling array employed in the inner refrigeration system belonged to an injection mold. Inner channels (2 mm in diameter) are manufactured through AM using DMLS

6.4 Cold Spray Low Pressure

Cold spray low pressure (CSLP) is a technology applied as part of the additive manufacturing process, which, in addition to manufacturing individual components, allows to carry out part repair operations [19]. It is due to this last characteristic that cold spray has been widely used by the industrial sector.

This technology is being applied in remanufacturing operations of metal components where operating temperatures below the melting temperature are required. Thus, it is ideal for depositing materials in which it is desired to retain a large part of their initial properties, as well as producing oxide-free deposits without an adverse influence on the substrate surface on which the deposit has been made (Fig. 6.10). In the manipulation of light alloys, such as those of aluminum and magnesium, low-pressure cold spray systems are widely used to carry out this kind of deposits, with the aim of applying a protective layer against corrosion on chemically compatible substrates.

6.4.1 Process and Equipment Used in CSLP

Cold spray is a technology characterized by the absence of high temperatures in its process. In such a process, small particles (~150 μm) are accelerated at a supersonic speed through the effect of a pressurized gas, which can be nitrogen or helium, and whose operating temperature reaches 800 °C. The particles projected at supersonic

6.4 Cold Spray Low Pressure

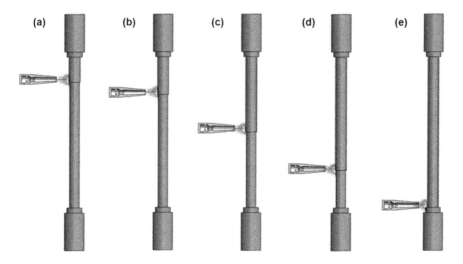

Fig. 6.10 Schematic representation of the remanufacturing process carried out on a metal piece. (**a–e**) A metal coat is deposited along the bar to restore functional working dimensions

speed create a coating whose porosity tends to decrease with the successive layers of applied material, allowing to reach a high hardness and density. The use of inert gases and temperatures below the melting temperature in the applied materials decreases the degree of reactivity and therefore the presence of oxides in the coating while maintaining the initial properties of the projected material, as well as its microstructure (Fig. 6.11).

The quality of the coating varies depending on the conditions in which the cold projection is carried out. The parameters to consider are type of material projected, as well as its physical and chemical properties, type of substrate, the contact surface roughness, and operating conditions of the equipment used. The issue concerning the deposit parameters is especially important in the case of light alloys. In the case of aluminum alloys, the temperature and pressure conditions reported in the literature are 300 °C and 0.6 MPa. If the pressure decreases to 0.5 MPa, the application temperature may vary between 27 and 626 °C, with a higher porosity depending on the temperature [20].

6.4.2 Powders Used in CSLP

Unlike laser sintering technologies for the manufacture of 3D structures, where metal particles with high uniformity in morphology and size are required, cold spray systems allow the manipulation of metal particles manufactured by other means (Fig. 6.12). The use of powder metallurgy and mechanical alloy for the production of metallic powders represents an economical alternative in the

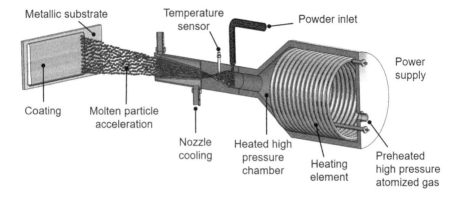

Fig. 6.11 Schematic representation of a cold spray array

Fig. 6.12 Secondary electron SEM micrographs of metal powder particles used in cold spray systems. Powder particles produced by (**a**) and (**b**) atomization and (**c**) and (**d**) mechanical synthesis (*Micrographs (a) and (b) courtesy: Cummins Inc*)

manufacture of alloys and advanced materials. In mechanical alloying and milling, the synthesis allows the use of elemental powders with different morphologies, as well as recycled materials obtained from machining processes. Current CS systems allow the handling of such particles in processes where it is necessary to apply coatings with high mechanical performance [21].

The characterization techniques involved in the analysis of metallic powders for AM allow us to study their morphology, size, and chemical composition. Commercial powders are commonly characterized by scanning electron microscopy (SEM), whose information is related to their uniformity in morphology and size. However, in those materials produced by mechanical synthesis, SEM provides additional information when their cross section is analized. A general idea of the degree of crystallinity in composites synthesized by mechanical synthesis can be directly observed through microstructural observations by SEM. This particular characteristic of the particles produced by this alternative route is a key factor in powders processed by CS. This is because an important fraction of the microstructure obtained from milling processes can be preserved after their deposition on substrates.

6.5 Conclusions

The potential of additive manufacturing is based on the expansion of its capabilities to produce parts with complex shapes and acceptable mechanical properties. Its innovation is moving fast, linked with new avenues to product design, using diverse materials, light alloys and composites. These advantages evidently have a considerable impact on the development of novel components used in industrial sectors as an emerging technology, with wide possibilities in the field of the aerospace, automotive, biomedical, electronics, and instrumentation industries. Although additive manufacturing is currently a technology away from a serial production process, its adaptability makes it a valuable tool for the production of customized and cost-effective prototypes and outstanding parts. Furthermore, the peculiarity of additive manufacturing to work complementary with high precision technologies, makes it an attractive methodology for the ad hoc remanufacturing of expensive parts, contributing to extending their performance, functionality, lifetime, and the reduction of costs instead of their total replacement.

References

1. Wong, K. V., & Hernandez, A. (2012). A review of additive manufacturing. *International Scholarly Research Notices, 2012*, 208760.
2. Sharma, G. S., et al. (2019). Influence of γ-alumina coating on surface properties of direct metal laser sintered 316L stainless steel. *Ceramics International, 45*(10), 13456–13463.
3. Vojnová, E. (2016). The benefits of a conforming cooling systems the molds in injection moulding process. *Procedia Engineering, 149*, 535–543.
4. Yuan, L., Ding, S., & Wen, C. (2019). Additive manufacturing technology for porous metal implant applications and triple minimal surface structures: A review. *Bioactive Materials, 4*, 56–70.
5. Gisario, A., et al. (2019). Metal additive manufacturing in the commercial aviation industry: A review. *Journal of Manufacturing Systems, 53*, 124–149.

6. Fathi, P., et al. (2018). A comparative study on corrosion and microstructure of direct metal laser sintered AlSi10Mg_200C and die cast A360.1 aluminum. *Journal of Materials Processing Technology, 259*, 1–14.
7. Qiu, X., et al. (2020). Influence of particulate morphology on microstructure and tribological properties of cold sprayed A380/Al$_2$O$_3$ composite coatings. *Journal of Materials Science and Technology, 44*, 9–18.
8. Han, Q., Setchi, R., & Evans, S. L. (2016). Synthesis and characterisation of advanced ball-milled Al-Al$_2$O$_3$ nanocomposites for selective laser melting. *Powder Technology, 297*, 183–192.
9. Anant Pidge, P., & Kumar, H. (2019). Additive manufacturing: A review on 3 D printing of metals and study of residual stress, buckling load capacity of strut members. *Materials Today: Proceedings, 21*, 1689–1694.
10. Liu, J., et al. (2018). Current and future trends in topology optimization for additive manufacturing. *Structural and Multidisciplinary Optimization, 57*(6), 2457–2483.
11. Wei, C., et al. (2018). 3D printing of multiple metallic materials via modified selective laser melting. *CIRP Annals, 67*(1), 245–248.
12. Peng, T., et al. (2018). Sustainability of additive manufacturing: An overview on its energy demand and environmental impact. *Additive Manufacturing, 21*, 694–704.
13. Levy, G. N. (2010). The role and future of the laser technology in the additive manufacturing environment. *Physics Procedia, 5*, 65–80.
14. Paz, R., et al. (2018). An analysis of key process parameters for hybrid manufacturing by material extrusion and CNC machining. *Bio-Design and Manufacturing, 1*(4), 237–244.
15. Zhang, X., et al. (2020). A review on energy, environment and economic assessment in remanufacturing based on life cycle assessment method. *Journal of Cleaner Production, 255*, 120160.
16. Olakanmi, E. O., Cochrane, R. F., & Dalgarno, K. W. (2015). A review on selective laser sintering/melting (SLS/SLM) of aluminium alloy powders: Processing, microstructure, and properties. *Progress in Materials Science, 74*, 401–477.
17. Deepika, S.S., B.T. Patil, & V.A. Shaikh. (2020). Plastic injection molded door handle cooling time reduction investigation using conformal cooling channels. *Materials Today: Proceedings*.
18. Jahan, S. A., & El-Mounayri, H. (2016). Optimal conformal cooling channels in 3D printed dies for plastic injection molding. *Procedia Manufacturing, 5*, 888–900.
19. Chen, C., et al. (2014). A review on remanufacture of dies and moulds. *Journal of Cleaner Production, 64*, 13–23.
20. Ajdelsztajn, L., et al. (2005). Cold spray deposition of nanocrystalline aluminum alloys. *Metallurgical and Materials Transactions A, 36*(3), 657–666.
21. Pialago, E. J. T., et al. (2015). Ternary Cu–CNT–AlN composite coatings consolidated by cold spray deposition of mechanically alloyed powders. *Journal of Alloys and Compounds, 650*, 199–209.

Chapter 7
Thermal Spray Coatings

Abstract Today, the manufacture of components with improved functionality and performance for industrial applications involves not only the selection of better and lighter materials but also the use of surface engineering to achieve superior resistance-to-weight ratio. For applications in aeronautical and aerospace sectors, friction, wear, temperature resistance, erosion, corrosion, adhesion, and surface finishing are some aspects of interest in research. The technique used for coating any substrate strictly depends on the type of material and the property that is required to improve. Nowadays, thermal spraying is a reliable and cost-efficient method to deposit thick coatings with a wide variety of feedstock materials and substrates. This chapter presents a review of the main thermal spray processes used to coat light alloys and shows experimental results of AlSiC and FeCrBSiNbW alloy coatings deposited on the 6061-T6 aluminum alloy. Splat formation and microstructure, including solid-liquid two-phase droplet impacting the coating deposition, are reviewed.

7.1 Introduction to Thermal Spray Processes

Among the most important light metals (aluminum, magnesium, titanium, and beryllium), aluminum alloys are by far the most used considering the worldwide production, by utilizing various combinations of their advantageous properties such as strength, lightness, corrosion resistance, recyclability, and formability. Besides, aluminum is being employed in an ever-increasing number of applications. However, it has a low melting point and very poor wear resistance [1]. In the same context, magnesium is approximately 35% lighter than aluminum, having good stiffness and damping capacity [2]. Magnesium alloys also have many advantages including low density, high specific strength and stiffness, excellent machinability, and good recyclability, which generate application value in many fields such as aerospace, automotive, and telecommunication industries [3]. However, limitations on magnesium alloy applications are related with their surface properties such as their poor corrosion and wear resistance, which are mainly attributed to their high reacting activity [4]. These disadvantages from two of the most important light alloys are sufficient reasons to use surface engineering, such as coatings, without losing the great advantage of their low weight compared to other metals. Thermal spray processes have been widely used in various industrial fields to enhance surface properties of different engineering parts. The application of coatings requires an understanding of the use of specific materials chosen, design or geometry of substrate, and the application process, which is usually dependent on energy consumption, raw material, and manufacturing cost.

The wear resistance, thermal resistance, mechanical properties, and corrosion resistance of engineering parts can be significantly improved by applying various coating materials to light metal surfaces through thermal spray processes [5]. Some features to use thermal spray coatings on light alloys are based on the thermal effects on the substrate during coating deposition, which can be minimized through controlling the relative motion of the traversing torch over the substrate, and a proper cooling to keep the substrate temperature below 150 °C. Some advantages of thermal spray are that coatings can be applied on parts machined to final shape without deteriorating the substrate microstructure and properties or causing deformation. Using thermal spray techniques, coatings can be deposited at a rapid deposition rate and thus at a low processing cost. They can be flexibly applied on the whole surface or on a local area of the part in a wide range of thicknesses.

The operating principle of all thermal spray processes is based on heating and melting a feedstock material, typically in the form of powder or wire, forming droplets, and/or semi-molten droplets. Then, a heater or a power source increase the temperature of particles, which are accelerated by a gas or plasma jet, to generate a high-velocity droplet stream acquiring sufficient kinetic energy (in the form of velocity) and thermal energy (in the form of high temperature) up to or over their melting points. During this stage, the chemical reaction between the droplet material and the flame atmosphere may lead to compositional changes of the particles from their starting composition and introduce new phases. Finally, the high-velocity droplets successively impact on the well-prepared rough surface of the substrate to adhere to it, followed by lateral flattening, rapid solidification, and cooling, which form a coating with a lamellar grain structure. These particles are mechanically or metallurgically bonded to the substrate. The process parameters to generate the droplets, including temperature, velocity, and size, which are determined by spray process parameters and conditions, influence the interaction of the spray particles with the spray flame [6]. Increasing atomizing gas velocity results in the formation of fine droplets with high velocity. Thus, highly dense coatings are produced. Figure 7.1 shows different thermal spray processes for different types of materials, which are classified based on the temperature and the particle velocity. Various types of materials, alloys, and composites in the form of powders, rods, and wires can be sprayed using the different thermal spray techniques [5]. The main advantages of using thermal spray processes include the use of materials and alloys with a higher melting point, possibility to coat large structural works in situ, a wide range of coating thicknesses, and lifetime economic processes, which produce coatings with long service life and less maintenance cost. Besides, chemical reactions, such as metal alloy oxidation during heating, may occur, which change chemical compositions and phases of the spray materials and may add additional functions to the coatings. However, spray coatings are not fully dense, with porosities of up to 20% depending on spray conditions. Further, their geometry presents two-dimensional

7.1 Introduction to Thermal Spray Processes

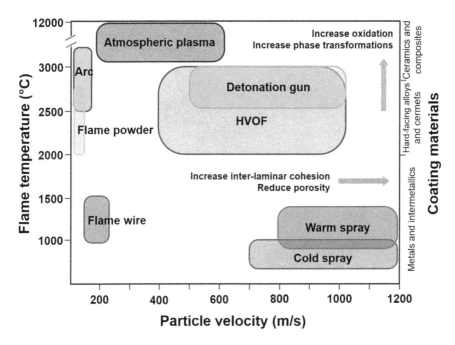

Fig. 7.1 Scheme of flame temperature range and particle velocity used in the operation of different thermal spray processes for coating manufacturing. Adapted from [8]

features that are different from the pores in bulk porous materials processed by other techniques such as powder metallurgy. The surface of a thermal spray coating is rough, so postsurface finishing may be necessary for some parts. Nevertheless, the pores in the coatings can act as a reservoir for lubricating oils to improve tribological performance. On the other hand, post-spray sealing may be necessary to achieve full corrosion protection [7].

In addition to the aforementioned aspects, substrate surface preparation and post-processing of the coating are also requirements. Almost all materials, including metal alloys, ceramics, plastics, and composites of those materials can be deposited to form coatings by thermal spraying. Powder materials are directly fed into a high temperature flame through a powder feeder to complete the first stage, while with wire or rod material spray droplets are created by atomizing the melting tip as it is fed into the flame. To completely or partially melt spray materials, various heat sources including gas flames, electrical arcs, and plasma jets are employed. The heating ability of a heat source is limited by its maximum temperature, which determines the types of materials that can be applied with a specific spray process, as will be explained in detail in Sect. 7.3.

Some physical and chemical phenomena are involved in the substrate/coating system, as shown in Fig. 7.2. These phenomena depend on the thermal spray process used, and some of these features lead to a good or bad coating, due to the high amount of variables that each process involves. In general, thermal spray coatings

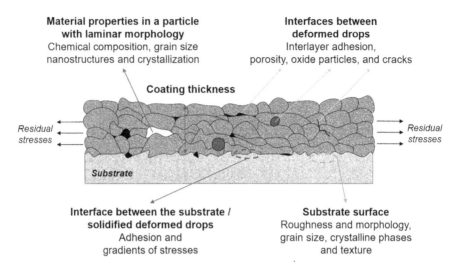

Fig. 7.2 Schematic general diagram of physical and chemical phenomena involved in the substrate/coating system obtained by thermal spray processes

can have between 100 μm and 3 mm in thickness, which is controllable depending on the property to be improved. For the thermal spray process, deposits generally have a lamellar structure if they are not posttreated. The main adhesion mechanism is mechanical, resulting in relatively low bonding strength. The microstructure (presence of pores, oxides, and/or unmelted particles) and hence the quality of the produced coating are a direct result of the used deposition technique. Some research is based on obtaining nanostructures, phase control, and crystallinity of the coating according to various chemical compositions and processes used [9, 10]. Residual stresses developed in thermal spray can be attributed to quenching, tension generated as the impacting splat contracts during cooling to the substrate temperature, and differential thermal contraction stresses, generated due to differences in thermal expansion as the splat/substrate system cools down to room temperature [11].

7.2 Thermal Spray Processes Used to Coat Light Alloys

In this section, the main thermal spray processes used to coat light alloys are described. Thermal spray methods, e.g., plasma, combustion flame, and high velocity oxy-fuel, are widely used to produce thermal coatings resistant to oxidation and wear for many industrial applications, such as aerospace, electronics, energy systems and oil, and gas [11]. The simplicity and flexibility of the process, in terms of flame temperatures chemistry and flow, make it suitable to melt any material, allowing the coating properties (e.g., density and chemistry) to be tuned over a wide range. Many of the properties of thermally sprayed coatings depend on their

7.2 Thermal Spray Processes Used to Coat Light Alloys   107

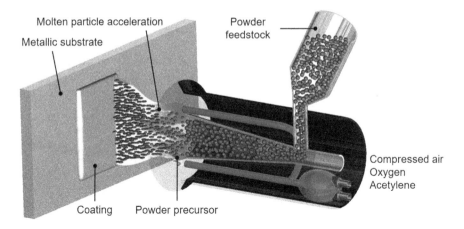

Fig. 7.3 Scheme of the combustion flame spray process

microstructure (porosity, oxides, particles overlapping), which in turn is dictated by the history of mechanical, thermal, and chemical phenomena involved during deposition. Cold spray is an emerging coating technology in which solid particles, at a high velocity and a temperature lower than their melting point, are used to deposit a coating through plastic deformation on impact. Cold spray is currently the most used process for coating light alloys such as aluminum alloys.

7.2.1 Combustion Flame Spray

The combustion flame spray (CFS) technique has some disadvantages compared with HVOF or plasma spray techniques. It produces coatings with modest quality including pores, oxides, cracks, and relatively poor adhesion to the substrate, having certain advantages such as being more economical, easier to implement, and more adaptable to a wide range of materials. It provides good wear and corrosion resistance, especially after thermal treatment. CFS is one of the most used techniques for the manufacture of tribological coatings for a wide variety of applications, including dry wear resistance or, in the presence of lubricants, wear resistance at high temperature, tribocorrosion conditions, and restoration of worn components [12]. Figure 7.3 shows the operation principle to obtain a coating by combustion flame spray process. In this process, the jet is a flame formed by the combustion between premixed oxygen and acetylene gases providing the energetic jet stream, and the injected feedstock is powder. Flame temperatures are up to 3100 °C, and particle velocities are typically in the range of 100–300 m/s depending on the alloy density, particle shape, and surface texture [13]. The particles soften in the flame, spread into splats upon impact with the substrate surface, transfer heat to the underlying substrate and solidify rapidly. The coating is thus built up from an agglomeration of solidified splats. During this process, the substrate receives heat from the solidifying splats, but it is also heated by the tail end of the flame, as the hot gas jet escapes from the impact area by flowing out sideways over the substrate. These two sources of heat can significantly raise the temperature of the substrate and may affect the properties of the temperature-sensitive light alloy. Al-SiC composites [7] and Ni-Si-B-Ag alloys [14] are materials that have been deposited on light alloys by this technique.

7.2.2 Arc Spray

In the twin arc spray process, the material precursor is fed into the system in the form of two cored wires, which melt within an electric arc. Figure 7.4a schematically shows the wire preparation to be used in this process. In the case of composite wire, it is formed by a metal alloy sheath and a powder that is placed inside such a sheath. The powder can be contain different particles sizes or complex chemical compositions, such as FeCrBSiNbW [9], FeBSiNbCr, and FeBSiCrNbMnY, which have high glass-forming ability, obtaining a composed coating microstructure formed by nanocrystalline phases in an amorphous matrix [15]. Compared with

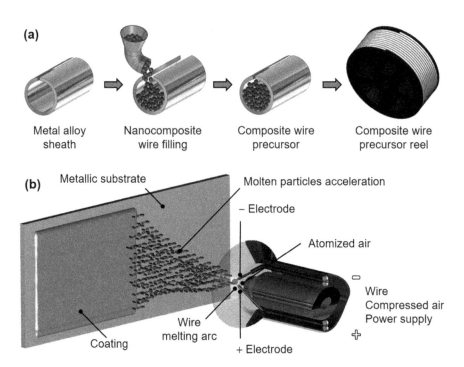

Fig. 7.4 (a) Wire components used in the arc spray process and (b) principle of the wire arc spray technique

other thermal spray processes, this process has some advantages due to its low cost and easy application. Further, the coatings can be synthesized successfully. Figure 7.4b shows the fundamentals to obtain a coating using the arc wire spray technique. The electrical arc spray process uses a direct current electric arc, struck between two consumable electrode wires, to melt the wires. A high-pressure gas jet is used to atomize the melts on the tips of the wires to create droplets with high velocity. The high spray rate of the arc process for different materials makes it the most cost-effective process. Arc spray is available to use with conductive filler materials in the form of wire, powder or powder-filled wire. The use of this process on light alloys is very limited; some materials such as aluminum [16, 17], Al alloys, Al-Zn alloys, and IN625 alloys [18] have been deposited on ferrous substrates.

7.2.3 High Velocity Oxy-Fuel

High velocity oxy-fuel (HVOF) process (Fig. 7.5) uses a mixture of fuel and oxygen, which is fed into a combustion chamber, where it is continuously ignited and combusted. The resulting hot gas emanates through a converging-diverging nozzle at a pressure close to 1 MPa and travels through a straight section. The fuels can be gases such as hydrogen, methane, propane, propylene, acetylene, and natural gas or liquids such as kerosene. The jet velocity at the exit of the barrel is usually over 1000 m/s, which exceeds the velocity of sound [8]. The powder material is fed into the jet at the feed ports, and the powder particles are heated and accelerated toward the substrate, where they impinge at high velocity to form a coating. WC-Co and WC-CoCr coatings have been applied onto three different light alloys substrates, Al-2024, Al-6082, and Al-7075 aluminum alloys, via the HVOF method [19]. Besides, Al and Al-SiC composites have been investigated for coating Mg alloys [2].

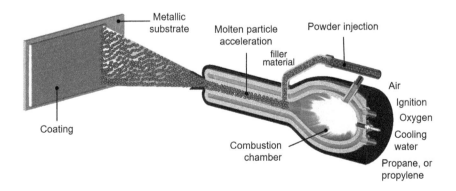

Fig. 7.5 Schematic operation principle of the HVOF process

7.2.4 Cold and Warm Spray

Cold spray has emerged as a promising surface engineering and additive manufacturing process for depositing metallic powder particles in the solid state. This is due to the very low process gas temperatures employed, which form dense metallic deposits with relatively low oxide content and superior interparticle adhesion. The particles are accelerated to supersonic velocities by compressed gas in a converging-diverging nozzle. Upon impact, high kinetic energy of the particle generates adiabatic shear instability, with which a successful bonding between particle and substrate or deposited coating is obtained. Good ductility makes aluminum and aluminum alloys convenient materials for cold spray applications, and, in general, successful bonding is generated via plastic deformation upon particle impact. However, sometimes cracks may develop in aluminum components, which have to be repaired by welding and conventional thermal spray techniques. Cold spray can be used as an alternative technique to repair such cracks in aluminum components [20, 21]. Cold-sprayed Al coatings can be applied to protect metal surfaces from atmospheric degradation, because a very thin and impervious oxide layer is formed on the aluminum surface. Instead of replacing the whole structure, repair and reclamation of the coating are possible, with the required protection coming from the sprayed aluminum structure. Hence, the cold spray process is highly attractive, enabling tremendous savings by becoming an inherent part of an integral manufacturing process for Al and its alloys. As shown in Fig. 7.6, the main elements of the cold spray setup are the spraying unit consisting of a prechamber and a supersonic nozzle, the powder feeder, the gas heater, and the source of compressed gas. Particles are accelerated to very high velocities by the carrier gas forced through a nozzle. Upon impact, solid particles with sufficient kinetic energy deform plastically and bond mechanically to the substrate to form a coating. The characteristic features of the cold spray process are much lower temperature (600 °C) and higher velocity of particles (>700 m/s) [22]. The critical velocity needed to form bonding depends on the material's

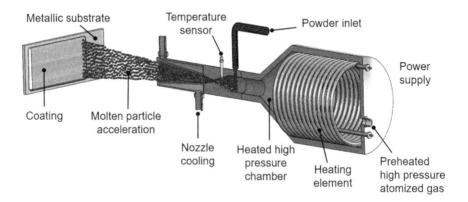

Fig. 7.6 Schematic operation principle of the cold spray process

properties, powder size, and temperature. Soft metals such as copper and aluminum are best suited for cold spray; however, the coating formation of other materials such as tungsten, tantalum, titanium, MCrAlY, and WC-Co by cold spray has been reported [21]. Nonetheless, the deposition efficiency tends to be low, and the window of process parameters and suitable powder size distribution tends to be narrow. Finer powders below 20 µm are typically used to accelerate the deposition rate. Concerning the processing gas, it is possible to accelerate powder particles to a significantly higher velocity using helium than using nitrogen, owing to its high velocity of sound; however, helium is much costlier, and its flow rate tends to be greater. To improve acceleration capability, nitrogen gas is heated to a higher temperature; the maximum gas temperature reached in the cold spray process appears to be approximately 900 °C [23]. It is important to emphasize that the key difference between cold spray and conventional thermal spray methods, from the physical viewpoint, is that in cold spray the coating is formed from particles in the solid state. The elimination of the harmful high-temperature effects on coatings and substrates offers significant advantages and new possibilities, such as avoiding oxidation and undesirable phases; retaining properties of initial particle materials; inducing low residual stresses; providing high density, high hardness, and cold-worked microstructure; spraying powders with a particle size <5–10 µm; working with highly dissimilar materials; feeding powder at a high rate, resulting in high productivity; depositing many materials at high deposition rates and efficiencies; heating the substrate minimally; and increasing operational safety because of the absence of high-temperature gas jets, radiation, and explosive gases. As a merit of heating the process gas, significant increases in the deposition efficiency and tensile strength of copper deposits were reported [22, 24]. It is possible to operate in the gas temperature range between HVOF and cold spray by mixing nitrogen with a combustion gas. A novel coating process called "warm spray" has been developed, in which coatings are formed by the high-velocity impact of solid powder particles heated to appropriate temperatures below the melting point of the powder material. The advantages of such a process are the following: (a) the critical velocity needed to form a coating can be significantly lowered by heating, (b) the degradation of feedstock powder such as oxidation can be significantly controlled compared with conventional thermal spray where powder is molten, and (c) various coating structures can be formed from porous to dense ones by controlling the temperature and velocity of the particles. In this process, the generated gas contains a significant amount of unreacted water vapor, hydrocarbons, and oxygen and therefore is not as clean as the gas used for the cold process. Therefore, the reaction of the powder with the gas is relatively slower because the powdered material remains unmelted during spraying.

The term warm spraying comes from the fact that this process operates in the intermediate temperature range between HVOF spray and cold spray; it is considered to have certain advantages and disadvantages over these processes. Cold spray has disadvantages inherent to the low temperature of the sprayed particles. As a consequence, there is no intimate bonding among the different splats, leading to a high porosity coating when using metal powders with low ductility [2]. In the case of ceramic particle addition to the coating to form metal matrix composites (MMCs), the porosity of the coating increases with increasing content of ceramic particles, due to lack of intimate bonding between the metal matrix and the ceramic reinforcement.

7.3 Spray Materials for Modification of Light Alloys

In general, materials stable at elevated temperature are suitable for thermal spray processing. Most metals, intermetallics, alloys, all forms of ceramics including oxides, borides, and silicides, cermets, and some polymers can be manufactured by one or more thermal spray processes [5]. Some typical materials used recently for modification of light alloy parts are listed in Table 7.1, in which typical industrial applications on light alloys and composites are shown. Aluminum (Al) and its alloy powders have promising coating properties when applied as a thermal spray on different types of substrates. As can be seen in this review, a considerable amount of research has been conducted on the coatings of Al and its alloys on different substrate materials, mostly using cold spray process.

Recent advancements in the manufacturing of lightweight structures have led to significant interest in magnesium (Mg) alloys because of their beneficial properties, such as high specific strength, stiffness, and low density [25]. As a lightweight material, Mg and its alloys are the primary candidates for the transportation vehicle lightweighting, as they can improve vehicle fuel economy and reduce greenhouse gas emissions. However, there are obstacles preventing further application of Mg alloys. The high electrochemical reactivity of Mg translates into lower corrosion resistance, which limits its application in humid conditions and in aqueous environments [26]. This is particularly a problem for applications involving structural components, since most of them experience cyclic loading in service. Therefore, the corrosion resistance and fatigue life of Mg alloys need to be significantly improved in order for transportation industries to take full advantage of the benefits of Mg lightweighting [27]. Many researchers have reported that the corrosion properties of Mg alloys can be improved by forming protective coatings, such as high velocity oxy-fuel (HVOF) coating [2], metal cladding [28], thermal spray coatings [29], and cold spray coating [25, 30–33] of different coating materials, which results in better microstructures and improved corrosion resistance.

As an example of the application of thermal spray techniques, rectangular-shaped plane plates of 6061-T6 aluminum alloy were thermally coated using two types of filler materials: AlSiC and FeCrBSiNbW coatings, which were applied by combustion flame spray process (CFS) and arc spray process (AS), respectively. 6061 aluminum alloy has excellent joining characteristics and good acceptance of applied coatings and combines relatively high strength, good workability, and high resistance to corrosion, in addition to being widely available. Some applications are aircraft fittings, couplings, marine fittings and hardware, electrical fittings and

7.3 Spray Materials for Modification of Light Alloys

Table 7.1 Different materials used for surface modification of light alloys and composites deposited by thermal spray processes

Substrate	Coating material	Applications	Process	References
Casting A357 alloy	A357 Al alloy	Aircraft and missile components like brackets, frames, motion transfer, and gearbox housings	Cold spray	[34]
6061-T6 Al	Al6061 Al alloy	Aircraft fittings, landing mats, truck bodies, and frames	Cold spray	[35]
AZ31B Mg alloy	AA7075 Al alloy	Trimming parts and a few structural parts, load-bearing parts	Cold spray	[25]
AA7075 Al alloy	AA7075 T73 wrought Al alloy	Aircraft fittings, gears and shafts, fuse parts, meter shafts and gears, missile parts, regulating valve parts, worn gears, keys, aircraft, aerospace, and defense applications	Cold spray	[31]
6061-T6 Al	Al-Cr-Mn-Co-Zr alloy	Components resistant to pitting corrosion in salt fog exposures	Cold spray	[36]
Al7075-T6	Al7075 Al alloy	Upper skins and spar caps of wings for transport aircraft	Cold spray	[37]
AZ31B Mg alloy	Pure Al	Aircraft fuselages, cell phone, and laptop cases	Cold spray	[32]
ZE41A Mg alloy	Al-SiC composite	Aerospace castings	HVOF	[38]
AZ91 Mg alloy	Al and Al/SiCp	Structural components for aerospace, automotive, and telecommunication industries	Flame spray and HVOF	[2]
Al-4%Cu alloy	WC-CoCr	Cutting tools, plungers, bearings, gears, steam turbine, hydro turbines	HVOF	[39]
Al2024, Al6082, and Al7075 alloys	WC-Co and WC-CoCr	Structures, gears and shafts, missile parts, orthopedic equipment and scientific instruments. Cranes, highly stressed applications, ore skips, bridges, and transport applications	HVOF	[19]
AZ31, AZ80, and AZ91D Mg/Al alloys	Al powder	Automotive industry to produce coatings on transmission and engine parts such as synchronizing rings, shift forks, and large volume of piston rings	Flame spray	[29]
AZ31, AZ80, and AZ91D alloys	Al/SiC composite	Automotive industry to produce coatings on transmission and engine parts	Flame spray	[7]
Al-Mg alloy (grade 5083)	Al enamel	Polymer injection molding machines	Flame spray	[40]
Cu	Ni-Si-B-Ag alloy	Structural materials	Flame spray	[14]

(continued)

Table 7.1 (continued)

Substrate	Coating material	Applications	Process	References
AZ91 Mg alloy	Al and AlCr$_6$Fe$_2$ powder alloy	Lightweight construction in the automotive and aerospace industries	Plasma spray	[41]
Steel	95Al5Mg and 94.4Al5Mg0.6Ti alloys	Marine structures such as the external parts of offshore wind power plants and large steel bridges	Plasma spray	[42]
Al6061, steel	SiC and Al/SiCp	Aerospace, automotive, structural parts	Plasma spray	[43]
Al-Si alloy	Cast iron powders	Engine cylinders and cast iron liners	Plasma spray	[44]
Plain carbon steel	Pure Al	Bridges, strobes, pipelines, and other large steel structures	Arc spray	[17]
SS304L	Al, ZnAl, and Inconel 625 alloy	Fuel storage from nuclear power plants	Arc spray and HVOF	[18]

connectors, hardware, hinge pins, brake and hydraulic pistons, appliance fittings, and valve parts.

In this exercise, for the combustion flame spray process, a Castolin-Eutectic DS8000 with a SSM40 modulus was used. For this equipment, a thermal power of 28 kW was supplied by mixing oxygen at 4 bar and 2000 L/h with acetylene at 0.7 bar and 1800 L/h. In this flame spray process, AlSiC alloy designated as 310NS (from Oerlikon) was used as precursor powder. On the other hand, the arc spraying was developed in a Praxair Tafa 8835 equipment with twin-wire arc torch, and a commercial Fe-based alloy designated as 140MXC nanocomposite wire (from Praxair Surface Technologies) was used as filler material. For both processes, samples of Al6061-T6 alloy with 10 cm × 10 cm and 0.5 cm in thickness were used as substrates. In order to enhance the adhesion of the coatings, the metallic surface was sand-blasted with Al$_2$O$_3$ grit 16 (particle size average 1092 µm) and then cleaned with acetone. The chemical compositions of these materials are shown in Table 7.2. The precursors were deposited on the cleaned Al-alloy substrates using the process parameters given in Table 7.3. In both processes, the spray parameters were selected based on preliminary experiments.

The aim of using two different processes was to obtain coatings with significantly different microstructures (porosity, splat geometry, and interface bonding). The key motive of this study was to understand how the mechanics and microstructure of thermal sprayed coatings are correlated with the coating tribological performance. The main challenge was to determine the relationship between the different types of coatings, based on processing variables according to each technique and obtain optimal conditions to fulfill the coatings requirements for aluminum alloy.

For the sample characterization, surface and cross sections of the coated specimens were wet ground through successive grades of SiC abrasive paper from 120 to 2000 grits, followed by diamond finishing to 0.1 µm. Field emission scanning

7.4 Microstructure of Coatings

Table 7.2 Chemical composition of substrate and spraying precursors

Material	Chemical composition (wt%)													
	Al	Cr	Cu	Mg	Mn	Si	Nb	Fe	Mn	C	Mo	W	B	Others
Al6061-T6 (substrate)	96.0	0.35	0.3	1.0	<0.15	0.6	–	<0.7	–	–	–	–	–	<0.55
310NS powder	64.0	–	–	–	–	7.0	–	–	–	22.0	–	–	–	–
140MXC composite wire	–	25.0	–	–	–	2.0	12.0	28.0	3.0	4.0	6.0	15.0	5.0	–

Table 7.3 Spraying process parameters

Process parameter	Combustion flame spray process (CFS)	Arc spray process (AS)
Voltage (V)	–	32
Current (A)	–	150
Air pressure primary (bar)	0.0	4.2
Air pressure secondary (bar)	0.0	2.8
Oxygen pressure (bar)	4.5	–
Acetylene pressure (bar)	1.0	–
Stand off-distance (cm)	20	20
Number of passes	2	2

electron microscopy (FE-SEM) and energy-dispersive spectroscopy (EDS) were carried out on a FEI Nova NanoSEM 200 microscope to observe surface and cross-section coating microstructures. The crystalline phases from both coatings and precursors were characterized by X-ray diffraction analysis using a Panalytical Empyrean diffractometer equipped with a CuK$_{\bar{u}FC}$ radiation ($\lambda = 1.5406$ Å), operated at 45 kV and 40 mA; an ultra-fast X'Celerator detector with a Bragg-Brentano geometry was used. XRD scans were performed with a step size of $0.016°$ s^{-1} and a dwell time per step of 40 s in the 2θ range of 20–130°. The phases were identified by comparing the experimental XRD patterns to standards compiled by the International Centre for Diffraction Data (ICDD), using High Score Software®.

7.4 Microstructure of Coatings

Figure 7.7 depicts SEM micrographs for comparing the morphology and chemical composition of the feedstock precursor powders. As can be seen in Fig. 7.7a, AlSiC powder (310NS) has spherical morphology with some kind of texture typically observed in powders manufactured by spray dry technology; an average particle size of 50 μm was measured. Figure 7.7b shows FeCrBSiNbW powder (140MXC), extracted from the wire as a precursor for the arc spray process, which presents an irregular blocky-type morphology, as well as a non-smooth surface; the particle size

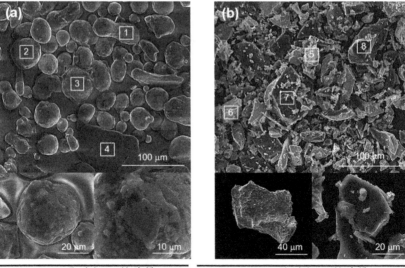

Fig. 7.7 SEM images and EDS analysis of feedstock powder showing (**a**) 310NS powder used in combustion flame spray process and (**b**) 140MXC powder used in arc spray process

is less than 100 μm. This last morphology was previously observed in amorphous powder synthesized by mechanical alloying [45]. The chemical composition of this powder varies depending on the type of each particle.

In the arc spray process, the parameters were selected according to previous author results [9]. For the combustion flame spray process, three gas pressure ratios were used to select the best. As a global view of surface and cross-section coatings, Fig. 7.8 presents the effect of this ratio on the overall coating microstructure obtained. As can be observed, the level of pores and oxide particles increase as the oxygen/acetylene gas pressure ratio is higher. This difference in gas pressure ratio also has an impact on the phase distribution. For example, when the PO_2/PAc ratio was 4.1, carbon phase was better distributed in the Al-Si matrix, and the presence of a certain level of pores was observed, but not too pronounced compared with the sample obtained using the PO_2/PAc ratio of 5.7. On the other hand, the PO_2/PAc ratio of 2.9 results in a carbon phase with bigger size, higher porosity, and unmelted particles.

The sample with $PO_2/PAc = 4.1$ was selected to analyze its microstructure in detail. Figures 7.9a–d show SEM images of the surface coating. In general, the surface is homogeneously coated. The low particle velocity characteristic of the flame thermal spray process ensures that some particles are not well spread at the time of impact on the substrate and, therefore, this phenomenon promotes the presence of

7.4 Microstructure of Coatings

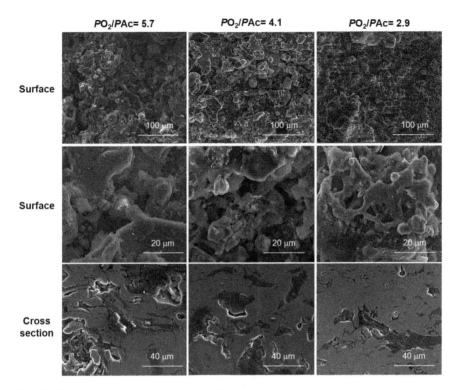

Fig. 7.8 Effect of gas pressure ratio used in the combustion flame spray process on the overall coating microstructures

pores in the coating, resulting in weakly bonded interfaces. As can be seen in Fig. 7.9c, the coating also presents a certain level of intra-particle cracks not too pronounced and a great quantity of oxides particles (Fig. 7.9d). According to the XRD results shown in Fig 7.9f, these oxides basically correspond to aluminum oxide. Figure 7.9e presents the general chemical composition of the coating. The XRD patterns in Fig. 7.9f also reveal the crystallinity of the precursor powder, which mostly correspond to the Al_9Si phase with a large amount of carbon and a smaller amount of silicon, as well as the phases found after the coating deposition. Carbon remains mostly in the elemental phase after being deposited as a coating, followed in quantity by the aluminum oxide, and then the $Al_{3.2}Si_{0.47}$ phase, very small amounts of SiC and negligible amount of the Al_4C_3 phase.

Figures 7.10a–c show BSE-SEM micrographs of the general appearance of the AlSiC cross-section. The average coating thickness was 434 μm, and, in general, the coating was perfectly well bonded to the substrate (Fig. 7.10a). According to the EDS results (Fig. 7.10d), the thermally sprayed coating is composed of an Al-Si matrix (gray color) with carbon phase well distributed between lamellar structures. The matrix plays, in most cases, a role in distributing stresses homogeneously in the coating. The carbon particles are easily identifiable by their different color (dark)

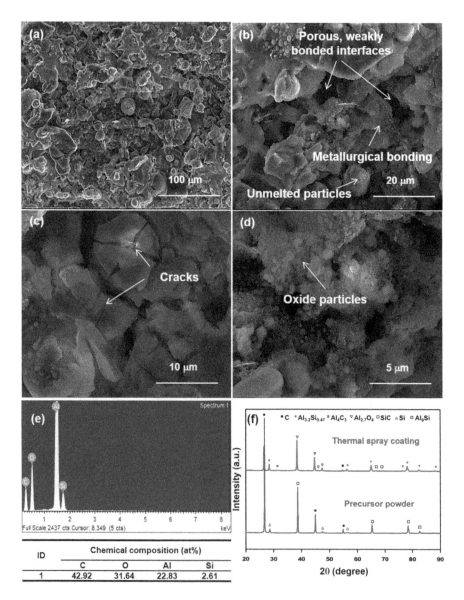

Fig. 7.9 SEM images of coating obtained by the combustion flame spray process: (**a**) and (**b**) surface morphology, (**c**) magnified intra-particle crack, (**d**) magnified oxide particles, (**e**) SEM-EDS elemental analysis of the surface, and (**f**) XRD patterns from precursor powder and surface coating

and irregular shape. During the coating process, the feedstock material begins to melt when the temperature reaches its melting point, which means that the melting point of AlSiC alloy is reached. However, some particles are not completely melted, and others reprecipitate with spherical morphology but with smaller size. Punctual

7.4 Microstructure of Coatings 119

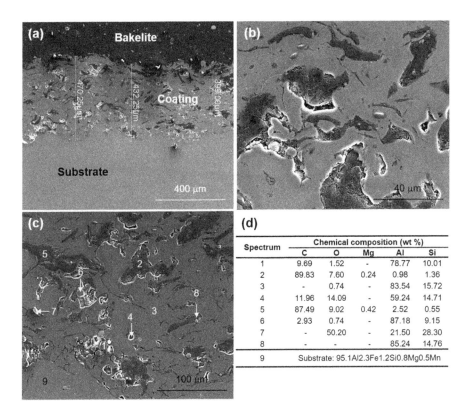

Fig. 7.10 (**a–c**) BSE-SEM images and (**d**) punctual EDS analysis taken from phases in the cross-section of the coating deposited by the combustion flame spray process (CFS)

EDS analyses were performed to identify each phase. The measurement spots are shown in Fig. 7.10c, and the results are summarized in Fig. 7.10d.

Figure 7.11 shows an elemental mapping analysis of the AlSiC coating. It can be noticed that the distribution of different phases within the coating has good overall homogeneity. As can be seen, the big particles correspond to the carbon phase, which are dispersed throughout the Al-Si matrix. The small particles well dispersed should be aluminum oxides formed during the deposition. The chemical composition is also illustrated in Fig. 7.11, from where the presence of oxides can be inferred, which is due to the relatively long residence time of the sprayed particles in the flame. Indeed, the particles are heated to a high temperature degree, and the chemical reaction with oxygen present in the environment is enhanced.

On the other hand, the surface morphology of the 140MXC coating is observed in Figs. 7.12a–d. As can be seen, the microstructure consists of individual splats with irregular morphology and little splashing, in addition to showing a good metallurgical bonding. This morphology suggests a good degree of deformation due to the high impact velocity of particles on the substrate. A low porosity is present, which was estimated as ~1.6% by image analysis. Some intra-splats cracks are also

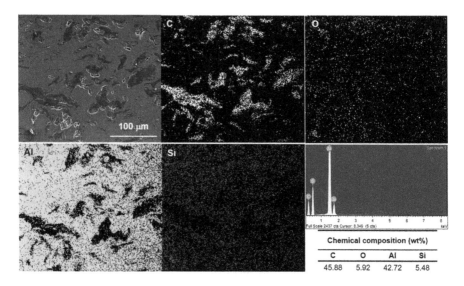

Fig. 7.11 SEM-EDS elemental mapping and SEM-EDS analysis spectrum taken from phases in the cross-section of the 310NS coating deposited by the combustion flame spray process (CFS)

visible, which are a consequence of the rapid cooling rate of the coating. Some spherical particles remain unmelted, while others melted and resolidified with a smaller size; other particles precipitate granularly with spherical morphologies. Figure 7.12e presents the average chemical composition of the surface coating, which was measured by EDS. Figure 7.12f shows a comparison of the XRD patterns from the precursor powder and 140MXC coating. The XRD pattern of the powder reveals a diffuse halo centered approximately at $2\theta = 43°$, which makes the presence of an amorphous structure evident. Likewise, several peaks corresponding to crystalline phases are observed; the phases identified are chromium boron carbide ($Cr_3B_{0.4}C_{1.6}$), boron carbide ($B_{13.3}C_{1.7}$), niobium carbide (Nb_6C_5), iron niobium ($Fe_{6.5}Nb_{6.5}$), tungsten (W), and a Fe-Cr solid solution. Some very small peaks between $2\theta = 80–100°$ were not possible to identify due to the complexity of the alloy chemical composition. It is inferred from the diffraction pattern that the precursor powder has a mixture of amorphous and crystalline phases. However, when this powder is projected by thermal spray coating, only two crystalline phases (Fe-Cr and boron carbide) are detected, which are within an amorphous phase. These results agree with previous studies [46] on similar Fe-based alloys deposited by HVOF, which suggest that the spray process promotes the increase of glass content on the coating compared to the precursor powder. The above is due to the fast quenching of molten particles through the action of flattening onto the substrate. According to these authors, the amorphous phase could be minimized by heat treatments given for long times at temperatures >800 °C. Nevertheless, compared with the HVOF process, the arc spray process can be considered as a potential technique for the synthesis of amorphous-crystalline composite coatings. In fact, the

7.4 Microstructure of Coatings

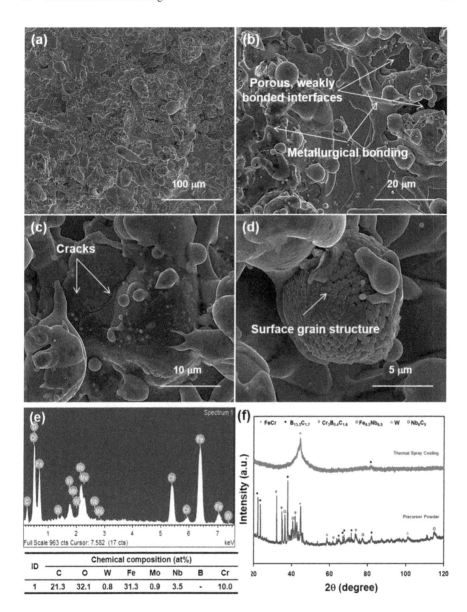

Fig. 7.12 SEM images of 140MXC coating obtained by arc spray process: (**a**) and (**b**) surface morphology, (**c**) magnified intra-particle crack, and (**d**) magnified sphere precipitate. (**e**) SEM-EDS elemental analysis of the surface and (**f**) XRD patterns from precursor powder and surface coating

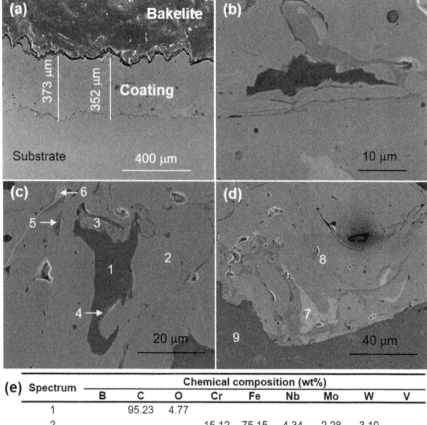

Fig. 7.13 (**a**–**d**) SEM images and (**e**) punctual SEM-EDS analysis taken from phases in the cross-section of the 140MXC coating deposited by the arc spray (AS) process

application of this technique has attracted increasing attention in recent years due to the high mechanical strength, strong corrosion resistance, and good soft magnetic properties that it provides to the projected materials [47].

In the same context, Figs. 7.13a–d show images of the cross-section 140MXC coating deposited by the arc spray process. The microstructure morphology is homogeneous, and the coating is well adhered to the substrate. The measured average thickness was 362 ± 10 µm. As can be seen, the coating presents a flattened lamellar structure with small cracks and low porosity derived from poorly stacked flat particles. The microstructure of the coating is mainly composed of four phases (Figs. 7.13c–d): dark particle (Zone 1), the gray lamellae (Zones 2 and 8), external shell with dark precipitates (Zone 5), and white precipitate (Zones 6 and 7). The dark particles and white precipitates, magnified in Figs. 7.13b–d, respectively, are heterogeneously dispersed in the whole microstructure. The EDS elemental chemical analysis of the different phases identified in both the cross-section of the coating and substrate is listed in Fig. 7.13e.

As a complement to the EDS elemental analysis presented in Fig. 7.13e, Fig. 7.14 presents EDS elemental mapping and average chemical composition taken in the cross-section of the 140MXC coating deposited by the arc spray process. The mapping was performed on a region presenting a precipitate in the interface boron carbide/matrix. As can be seen, Fe and Cr are observed mainly in the matrix, and the rounded dark phase is related to boron carbide. Nb was found dissolved and well dispersed within the matrix; it appeared in Cr-rich regions. W was also found well dispersed within the matrix without forming precipitates. Mo was present in a white phase surrounding the carbide phases. Oxygen was found dispersed and directly related to porosity. Carbon was also found in acicular dark particles with different sizes, where no other elements were detected. The boron carbide growth in the crystalline-matrix interface can be attributed to partial crystallization by heterogeneous nucleation. The presence of boron carbides at the particle-matrix interface reveals that the mechanism of crystallization is similar to that developed in the matrix. Previous studies have reported that these interfaces were formed through adding reinforcement phases like Mo, Nb, and Ta on Zr-based bulk amorphous alloy [47]. Therefore, some advantages of producing in situ composite coatings by thermal arc spray are the fact that the contents of amorphous and crystalline phases can be controlled during the process. It is worth noting that in these coatings the interfaces between particles can represent a better mechanical and wear performance due to lower crack propagation through these regions.

7.5 Conclusions

The comprehensive understanding of the features of the different thermal spray coating processes contributes to the selection of the most suitable process for a particular material and application in light alloys and composites. Due to the range of temperatures that cold spray and combustion flame spray processes can be handled, they are the most used processes for manufacturing light alloys and composites. According to the experimentally investigated processes in this chapter, arc spray is a cost-effective process in which only alloy wires or cored wires are applicable. Alternatively, the combustion flame spray process is a very simplified process that offers advantages in its portability, making it available for in site industrial

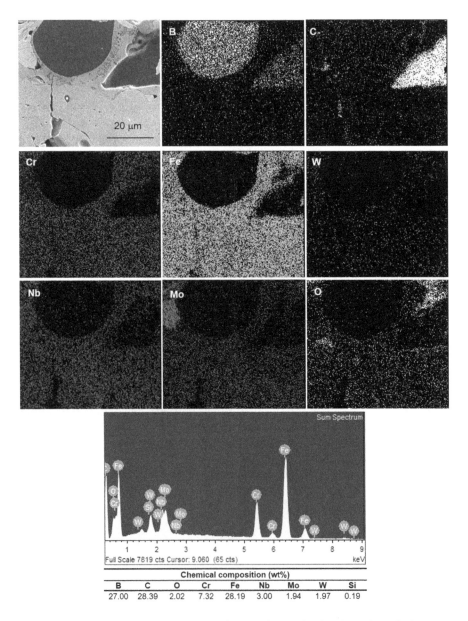

Fig. 7.14 SEM-EDS elemental mapping and SEM-EDS analysis taken from phases in the cross-section of the 140MXC coating deposited by the arc spray (AS) process

applications. Regarding the arc spray coating process, in our case study, the microstructure consisted of a Fe- and Cr-rich amorphous matrix with boron carbides and a tungsten phase randomly distributed, as well as molybdenum precipitated at the boron carbide-amorphous matrix interface. On the other hand, the as-sprayed AlSiC coating revealed a high level of porosity, which could negatively affect the mechanical properties; nonetheless, a better contact between the deposited splats and improved bonding at the substrate-coating interface was observed. The formation of diffusion layers, degradation of particles, and the formation of aluminum carbides were not observed. The features inherent to thermal spray processes and coating microstructures can be considered as both advantages and disadvantages, according to the application. Therefore, optimization of spray conditions and coating microstructures must be made according to the applications on components manufactured from light alloys and composites.

References

1. Polmear, I., et al. (2017). The light metals. In *Light alloys. Metallurgy of the light metals*. Amsterdam: Elsevier, Butterworth-Heinemann.
2. Lopez, A. J., et al. (2015). Optimisation of the high velocity oxygen fuel (HVOF) parameters to produce effective corrosion control coatings on AZ91 magnesium alloy. *Materials and Corrosion, 66*(5), 423–433.
3. Bonnah, R. C., Fu, Y., & Hao, H. (2019). Microstructure and mechanical properties of AZ91 magnesium alloy with minor additions of Sm, Si and Ca elements. *China Foundry, 16*(5), 319–325.
4. Lu, X., et al. (2019). Investigation of protective performance of a Mg-rich primer containing aluminum tri-polyphosphate on AZ91D magnesium alloy in simulated acid rain. *Coatings, 9*(10), 649.
5. Li, C. J. (2010). Thermal spraying of light alloys. In H. Dong (Ed.), *Surface engineering of light alloys*. Cambridge, UK: Woodhead Publishing Limited.
6. Christopher, C., & Berndt, W. J. L. (2004). In J. R. Davis (Ed.), *Handbook of thermal spray technology*. Materials Park: ASM International.
7. Arrabal, R., et al. (2010). Al/SiC thermal spray coatings for corrosion protection of Mg–Al alloys in humid and saline environments. *Surface and Coatings Technology, 204*(16–17), 2767–2774.
8. Oksa, M., et al. (2011). Optimization and characterization of high velocity oxy-fuel sprayed coatings: Techniques, materials, and applications. *Coatings, 1*(1), 17–52.
9. Arizmendi-Morquecho, A., et al. (2014). Microstructural characterization and wear properties of Fe-based amorphous-crystalline coating deposited by twin wire arc spraying. *Advances in Materials Science and Engineering, 2014*, 1–11.
10. Yin, S., et al. (2018). Cold-sprayed metal coatings with nanostructure. *Advances in Materials Science and Engineering, 2018*, 1–19.
11. Fanicchia, F., et al. (2018). Residual stress and adhesion of thermal spray coatings: Microscopic view by solidification and crystallisation analysis in the epitaxial CoNiCrAlY single splat. *Materials & Design, 153*, 36–46.
12. Rachidi, R., El Kihel, B., & Delaunois, F. (2019). Microstructure and mechanical characterization of NiCrBSi alloy and NiCrBSi-WC composite coatings produced by flame spraying. *Materials Science and Engineering B, 241*, 13–21.
13. Fanicchia, F., et al. (2017). Combustion flame spray of CoNiCrAlY & YSZ coatings. *Surface and Coatings Technology, 315*, 546–557.
14. Ziewiec, K., et al. (2017). Thermal characteristics and amorphization in plasma spray deposition of Ni-Si-B-Ag alloy. *Journal of Alloys and Compounds, 710*, 685–691.
15. Cheng, J. B., Wang, Z. H., & Xu, B. S. (2012). Wear and corrosion behaviors of FeCrBSiNbW amorphous/nanocrystalline coating prepared by arc spraying process. *Journal of Thermal Spray Technology, 21*(5), 1025–1031.

16. Adamiak, M., et al. (2018). The properties of arc-sprayed aluminum coatings on armor-grade steel. *Metals, 8*(2), 142.
17. Lee, H.-S., et al. (2016). Corrosion resistance properties of aluminum coating applied by arc thermal metal spray in SAE J2334 solution with exposure periods. *Metals, 6*(3), 55.
18. Yung, T.-Y., et al. (2019). Thermal spray coatings of Al, ZnAl and inconel 625 alloys on SS304L for anti-saline corrosion. *Coatings, 9*(1), 32.
19. Varol Özkavak, H., et al. (2019). Comparison of wear properties of HVOF sprayed WC-Co and WC-CoCr coatings on Al alloys. *Materials Research Express, 6*(9), 096554.
20. Cavaliere, P., & Silvello, A. (2017). Crack repair in aerospace aluminum alloy panels by cold spray. *Journal of Thermal Spray Technology, 26*(4), 661–670.
21. Pathak, S., & Saha, G. (2017). Development of sustainable cold spray coatings and 3D additive manufacturing components for repair/manufacturing applications: A critical review. *Coatings, 7*(8), 122.
22. Coddet, P., et al. (2016). On the mechanical and electrical properties of copper-silver and copper-silver-zirconium alloys deposits manufactured by cold spray. *Materials Science and Engineering A, 662*, 72–79.
23. Cavaliere, P. (2018). *Cold-spray coatings. Recent trends and future perspectives*. Berlin: Springer Nature.
24. Chang, Y., et al. (2020). Microstructure and properties of Cu–Cr coatings deposited by cold spraying. *Vacuum, 171*, 109032.
25. Dayani, S. B., et al. (2018). The impact of AA7075 cold spray coating on the fatigue life of AZ31B cast alloy. *Surface and Coatings Technology, 337*, 150–158.
26. Joost, W. J., & Krajewski, P. E. (2017). Towards magnesium alloys for high-volume automotive applications. *Scripta Materialia, 128*, 107–112.
27. Jahed, H., & Albinmousa, J. (2014). Multiaxial behaviour of wrought magnesium alloys – A review and suitability of energy-based fatigue life model. *Theoretical and Applied Fracture Mechanics, 73*, 97–108.
28. Liu, J., et al. (2017). Research and development status of laser cladding on magnesium alloys: A review. *Optics and Lasers in Engineering, 93*, 195–210.
29. Pardo, A., et al. (2009). Corrosion protection of Mg/Al alloys by thermal sprayed aluminium coatings. *Applied Surface Science, 255*(15), 6968–6977.
30. Diab, M., Pang, X., & Jahed, H. (2017). The effect of pure aluminum cold spray coating on corrosion and corrosion fatigue of magnesium (3% Al-1% Zn) extrusion. *Surface and Coatings Technology, 309*, 423–435.
31. Rokni, M. R., et al. (2015). An investigation into microstructure and mechanical properties of cold sprayed 7075 Al deposition. *Materials Science and Engineering A, 625*, 19–27.
32. Shayegan, G., et al. (2014). Residual stress induced by cold spray coating of magnesium AZ31B extrusion. *Materials & Design, 60*, 72–84.
33. Wang, Q., et al. (2014). High resolution microstructure characterization of the interface between cold sprayed Al coating and Mg alloy substrate. *Applied Surface Science, 289*, 366–369.
34. Petráčková, K., Kondás, J., & Guagliano, M. (2017). Mechanical performance of cold-sprayed A357 aluminum alloy coatings for repair and additive manufacturing. *Journal of Thermal Spray Technology, 26*(8), 1888–1897.
35. Nautiyal, P., et al. (2018). In-situ mechanical investigation of the deformation of splat interfaces in cold-sprayed aluminum alloy. *Materials Science and Engineering A, 737*, 297–309.
36. Watson, T. J., et al. (2017). Cold spray deposition of an icosahedral-phase-strengthened aluminum alloy coating. *Surface and Coatings Technology, 324*, 57–63.
37. Sabard, A., et al. (2020). Cold spray deposition of solution heat treated, artificially aged and naturally aged Al 7075 powder. *Surface and Coatings Technology, 385*, 125367.
38. López, A. J., et al. (2013). Influence of high velocity oxygen-fuel spraying parameters on the wear resistance of Al–SiC composite coatings deposited on ZE41A magnesium alloy. *Materials & Design, 43*, 144–152.

39. Koutsomichalis, A., Vardavoulias, M., & Vaxevanidis, N. (2017). HVOF sprayed WC-CoCr coatings on aluminum: Tensile and tribological properties. *IOP Conference Series: Materials Science and Engineering, 174*, 012062.
40. Bao, Y., et al. (2013). Thermal-spray deposition of enamel on aluminium alloys. *Surface and Coatings Technology, 232*, 150–158.
41. Kubatík, T. F., et al. (2017). Mechanical properties of plasma-sprayed layers of aluminium and aluminium alloy on AZ 91. *Materiali in Tehnologije, 51*(2), 323–327.
42. Yoshio Shin, Y. O., Morimoto, T., Kumai, T., & Yanagida, A. (2016). Formation of nano-microstructured aluminum alloy film using thermal spray gun with ultra rapid cooling. *Materials Transactions, 57*(4), 488–493.
43. Tailor, S., Mohanty, R. M., & Soni, P. R. (2013). A review on plasma sprayed Al-SiC composite coatings. *Journal of Materials Science & Surface Engineering, 1*(1), 15–22.
44. Wang, Q., et al. (2019). Bonding and wear behaviors of supersonic plasma sprayed Fe-based coatings on Al-Si alloy substrate. *Surface and Coatings Technology, 367*, 288–301.
45. Movahedi, B., Enayati, M. H., & Wong, C. C. (2010). Structural and thermal behavior of Fe-Cr-Mo-P-B-C-Si amorphous and nanocrystalline HVOF coatings. *Journal of Thermal Spray Technology, 19*(5), 1093–1099.
46. Chokethawai, K., McCartney, D. G., & Shipway, P. H. (2009). Microstructure evolution and thermal stability of an Fe-based amorphous alloy powder and thermally sprayed coatings. *Journal of Alloys and Compounds, 480*(2), 351–359.
47. Li, J., et al. (2018). Synthesis of bulk amorphous alloy from Fe-base powders by explosive consolidation. *Metals, 8*(9), 727.

Chapter 8
Characterization Techniques

Abstract This chapter focuses on the main characterization techniques used for the morphological, microstructural, and structural analysis, as well as for the mechanical behavior analysis of light alloys and composites. There are different characterization techniques available, ranging from the most common and accessible ones to the most sophisticated and, sometimes, not available to everyone. A brief description of all these techniques is provided, presenting some examples of each one to show their capabilities. Additionally, the chapter is intended to inform the reader about the nuances and limitations during the characterization of metal matrix composites. The understanding of the microstructural features, such as the distribution of the reinforcement within the matrix and the interfacial reaction between them, as well as of the mechanical, thermal, and corrosion behavior is necessary during the production of metal matrix composites.

8.1 Introduction to Characterization Techniques

Different characterization techniques are available for the analysis of light alloys and composites, ranging from the most common and accessible techniques to the most sophisticated and, sometimes, not available to everyone. When some research or work in the industry is initiated and subsequently monitored, it is important to ensure and control that the starting materials, by-products, and final products have the suitable characteristics and properties. The chemical composition of materials can be determined by techniques like inductively coupled plasma, atomic absorption spectrophotometry, and CHNS-O elemental analysis. The main techniques of thermal analysis, namely, differential scanning calorimetry, differential thermal analysis, and thermogravimetric analysis, are used to quantify the change in the properties as a function of the temperature. The density of light alloys and composites can be measured by the Archimedes' method, pycnometry, or even computed tomography scanning. The microstructural study of samples can be performed from simple and economical techniques like optical microscopy to complex and expensive techniques like scanning electron microscopy, transmission electron microscopy, high-resolution transmission electron microscopy, and scanning transmission electron microscopy. Elemental chemical analysis of alloys and composites can be achieved by energy-dispersive spectroscopy and electron energy loss spectroscopy, in addition to identify the matrix-reinforcement chemical interaction. The structural analysis may be made by X-ray diffraction and synchrotron high-energy X-ray diffraction. Quantitative analysis and information on the chemical state of the material surface can be achieved by X-ray photoelectron spectroscopy. Vibrational, rotational, and other frequency modes of molecules, structural fingerprint, the electronic nature, and density of defects may be observed by Raman spectroscopy. The mechanical properties can be estimated by several techniques and at different levels: bulk, micro-, and nanoscales. Examples of these techniques are tensile, compression, flexural, wear, hardness, and nanoindentation tests. The next sections describe the use of some of these techniques for the analysis of light alloys and composites, presenting some examples of each one to see their capabilities.

8.2 Chemical Analysis

There are several techniques to determine the chemical composition, both qualitatively and quantitatively, of materials. They range from wet chemical methods to modern instrumental methods. The former cover classical methods that involve the analysis in the liquid phase, with the main use of laboratory glassware. The latter use sophisticated equipment, being some techniques the following: atomic absorption spectrometry (AAS), flame atomic absorption spectrometry (FAAS), inductively coupled plasma optical emission spectrometry (ICP-OES), inductively coupled plasma mass spectrometry (ICP-MS), and CHNS-O elemental analyzer. These techniques cover a wide range of concentrations for various applications. As an example of the techniques mentioned above, the ICP-OES technique will be briefly explained, and results of the chemical composition for a 6061 aluminum will be given.

The ICP-OES is an analytical technique based on the principles of atomic spectroscopy, which is useful to determine the elemental composition in materials of different nature; its concentration range is from major to trace (parts per billion) [1]. In theory, all elements of the periodic table could be determined by techniques based on ICP, but, in practice, there are restrictions in the analysis with elements artificially produced, inert gases and oxygen, due to their physical and spectral properties. Nowadays, there are automated instruments with multiple detectors capable of simultaneously determining 40 or more elements in a sample in less than 1 min. A disadvantage of ICP-OES, as all as wet chemical and instrumental methods, is that it is a destructive technique. Figure 8.1 shows a photograph of the equipment employed to determine the chemical composition of a 6061 aluminum alloy and presents the results of this analysis, which fulfill those reported in the literature [2].

8.4 Density Measurement

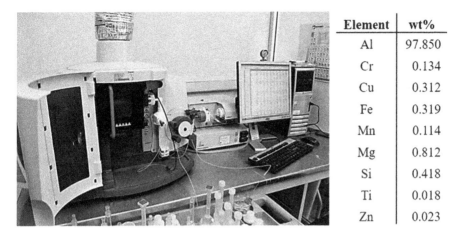

Element	wt%
Al	97.850
Cr	0.134
Cu	0.312
Fe	0.319
Mn	0.114
Mg	0.812
Si	0.418
Ti	0.018
Zn	0.023

Fig. 8.1 Thermo Scientific TM iCap 6000 Series ICp Spectrometer and results of the 6061 aluminum alloy analysis

8.3 Thermal Analysis

Thermal analysis refers to a set of techniques in which physical and chemical properties are measured as a function of the temperature. Transformation temperatures, fraction solids or fraction transforming, and latent heats, among others, can be determined [3]. The most used thermal analysis techniques are differential scanning calorimetry (DSC), differential thermal analysis (DTA), and thermogravimetric analysis (TGA).

As an example of the use of thermal analysis techniques, Fig. 8.2 presents a simultaneous TGA-DSC analysis achieved for a $Ti_{75}Mg_{25}$ sample synthesized by high-pressure torsion (HPT) straining. The TGA analysis (green curve) shows a weight loss of ~9%. The DSC analysis (blue curve) shows an endothermic peak at ~530 °C corresponding to the oxidation of Mg, and a peak at ~650 °C matching to the melting point of Mg.

8.4 Density Measurement

Density is an elementary physical property of materials, which can be measured by different methods and accuracies. Taking into account that one of the objectives in the development of light alloys and composites is to obtain the best weight-to-strength ratio, density measurement is essential. In the case of composites, apart from giving information on the decrease or increase in mass with the addition of a reinforcing material, a density result may suggest the presence of pores caused by the formation of reinforcement clusters. There are several possibilities for measuring the density of a part. The methods can be from the traditional and economic

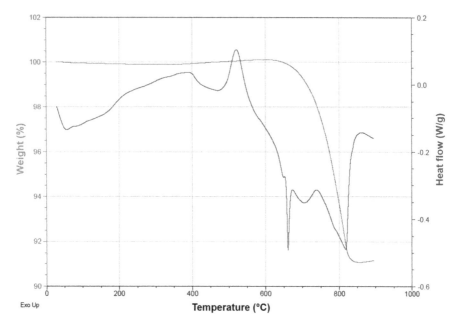

Fig. 8.2 TGA-DSC analysis of a Ti$_{75}$Mg$_{25}$ alloy processed by HPT

Archimedes' method, through the microstructural analysis of a sample cross section, to the most sophisticated as X-ray or neutron imaging [4, 5]. The choice of the method depends on different aspects, such as the availability, accuracy, and reproducibility that it can offer. In this chapter, the Archimedes', pycnometer and computed tomography scanning methods will be covered.

8.4.1 Archimedes' Method

The Archimedes' principle establishes that a solid immersed in a liquid is subjected to the force of buoyancy, whose value is the same as that of the weight of the liquid displaced by the volume of the solid [6]. The Archimedes' method shows a very high accuracy.

In this exercise, the density of the samples was measured using a Sartorius analytical balance (precision of 0.01 mg), to which a Sartorius YDK 01 density determination kit was adapted. The measurements were conducted with deionized water at 25 °C. The measurements were made in triplicate, using acetone for drying the samples before each measurement. The density was calculated by Eq. 8.1 [6].

$$\rho = \frac{w_a \left[\rho_w - 0.0012 \text{g}/\text{cm}^3 \right]}{0.99707 \cdot (w_a - w_l)} + 0.0012 \text{g}/\text{cm}^3 \quad (8.1)$$

8.4 Density Measurement

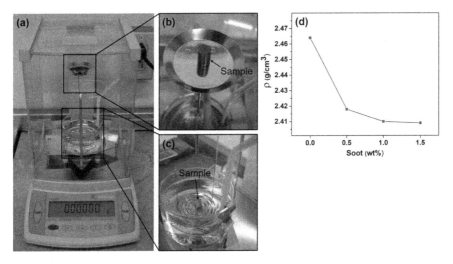

Fig. 8.3 Setup of the (**a–c**) Archimedes' method and (**d**) density of the Al6061 alloy reinforced with different fullerene soot contents

where ρ is the sample density, w_a is the sample weight in air, w_l is the sample weight in water, and ρ_w is the density of water at 25 °C (0.99707 g/cm³).

Figure 8.3a shows the accessories for measuring density, and Figs. 8.3b–c shows a sample being weighed in air and in water, respectively. Figure 8.3d presents the results of density measurements for the Al6061-fullerene soot composites synthesized by mechanical milling and conventional sintering. As can be seen, the addition of fullerene soot into de Al6061 matrix diminishes the density of composites, which can be due to the fact that the carbon has low tendency to make bonds with the aluminum leading to a weak network [7].

8.4.2 Pycnometry

A pycnometer can be used to measure the density of solids by employing the Archimedes' principle of fluid displacement. It consists of a jar of known volume, which can be equipped with a thermometer. Figure 8.4 shows a pycnometer with a 25-ml jar and thermometer.

For the density determination, the weight of the object (m_o) is measured. Then the weight of the pycnometer together with the inserted object ($m_p + m_o$) is measured. After that, distilled water is added, and the weight of the all ($m_p + m_o + m_{H2O}$) is measured. The difference between the latter and the former weights is the weight of water (m_{H2O}). The volume of the added water (V_{H2O}) can be obtained by Eq. 8.2.

Fig. 8.4 Pycnometer for measuring the density of samples

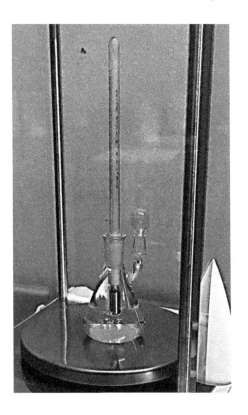

$$V_{H2O} = \frac{m_{H2O}}{\rho_{H2O}} \qquad (8.2)$$

The volume of the measured object (V_o) is the difference between the volume of water that fills the empty pycnometer (V) and the volume of water (V_{H2O}), according to Eq. 8.3). Note that m'_{H2O} is the weight of water that fills the empty pycnometer. Finally, the density of the object (ρ_o) is calculated by Eq. 8.4.

$$V_o = V - V_{H2O} = \frac{m'_{H2O} - m_{H2O}}{\rho_{H2O}} \qquad (8.3)$$

$$\rho_o = \frac{m_o}{V_o} \qquad (8.4)$$

8.4 Density Measurement

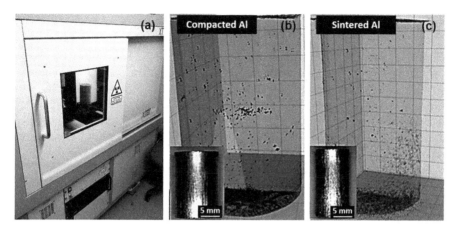

Fig. 8.5 (**a**) XT H 250 Nikon tomography system. CT images of (**b**) cold compacted and (**c**) sintered pure aluminum (*Images courtesy: Dr. Caleb Carreño-Gallardo*)

8.4.3 Computed Tomography Scanning

The computed tomography (CT) scanning is a computer-aided tomography process using X-ray irradiation to produce three-dimensional representations of a scanned object [8]. Numerous radiography projections are taken through a full 360° rotation, which are then back projected to reconstruct the 3D volume of the object. The technique is used in many areas of industry mainly for the internal inspection of components to detect and quantify porosity or defects, as well as to visualize their 3D distribution. However, it is also possible to obtain accurate volumetric measurements from parts with which, knowing their mass, their density can be determined [9].

For the case presented here, a XT H 250 Nikon tomography system (Fig. 8.5a) was used to examine the pore size distribution on pure aluminum. The measurements were achieved in a cold compacted Al sample and a sintered Al sample. The samples were placed on a rotating turntable between the X-ray source and the detector. As X-rays impinge on the sample, they are either attenuated or pass through the sample resulting in grayscale radiographs. The porosity distribution of the cold compacted sample and the sintered sample is shown in Figs. 8.5b–c, respectively. For a better visualization of the porosity, the matrix phase was removed from the entire volume. The images allow to see quantitatively that the porosity of the sintered sample is smaller and better distributed than in the compacted sample.

8.5 Optical Microscopy

Optical microscopy (OM), also known as light microscopy, commonly uses visible light and a lens system to generate magnified images of a sample. OM is a nondestructive and real-time imaging technique, being one of the most powerful and versatile investigation techniques in material sciences. It is the simplest technique for the superficial and morphological characterization, which allows to determine fairly quickly several features about the microstructure of analyzed samples. For instance, it is possible to identify the phases present in light alloys and composites, as well as to quantify a macroreinforcement in metal matrix composites.

Figure 8.6 shows images of different alloys and composites, where different microstructural features can be seen. Examples of these features are the grain morphology and size, the microstructural changes after a rolling or annealing process, the presence of twins, and diffusion.

8.6 X-Ray Diffraction

X-ray diffraction (XRD) is an important and versatile tool used in materials science for the characterization of materials. It is a rapid analytical technique mainly used for the phase identification of a crystalline material, providing information on the unit cell [10]. XRD is able to give immediate information on the phase composition and crystalline features after synthesizing and processing a sample. In the case of composites, it can detect the reaction between the matrix and the reinforcement, the fiber orientation, the effect of the reinforcement on the grain size, and the dislocation density. The technique is very sensitive to composition, but additions around 1% are difficult to identify [10].

As examples of the XRD capabilities, Fig. 8.7 shows diffraction patterns of the AZ31B Mg alloy reinforced with MWCNTs, on which a plasma electrolytic oxide (PEO) coating was applied. The patterns evidence the interaction between SiO_2 and MgO, allowing the formation of the Mg_2SiO_4 phase, whose formation rate increases with the increasing concentration of the $Na_2SiO_3.5H_2O$ electrolyte solution during the PEO coating. Additionally, the patterns reveal the presence of Mg, MWCNTs, and the formation of the MgO and $Mg_{17}Al_{12}$ phases.

XRD can be used for studying the crystal orientation and fiber orientation in composites. For example, the analysis of randomly oriented MWCNTs will result in a large peak at $2\theta = 26°$ corresponding to the (1 0 0) plane. Several low-intensity peaks corresponding to the (1 0 0) plane at $2\theta = 42.4°$ and the (1 1 0) plane at $2\theta = 77.7°$ may also be present indicating the MWCNTs presence. Qualitative information on the alignment of the MWCNTs can be made from the relative intensities of the different peaks [11].

Another use of XRD is in discerning the crystalline/grain size from the broadening of the peaks using, for instance, the simplified Scherrer formula (Eq. 8.5).

8.6 X-Ray Diffraction

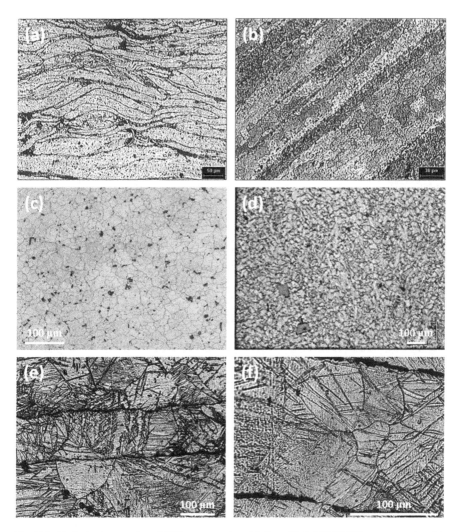

Fig. 8.6 Optical micrographs of different samples. (**a**) As-milled Al2024-NP$_{Ag-C}$ composite. (**b**) As-extruded Al2024-NP$_{Ag-C}$ composite. (**c**) Al6061-fullerene soot composite. (**d**) Hot rolled AZ31B Mg alloy. (**e–f**) AZ31B-MWCNTs composite

$$d = \frac{0.9\lambda}{B\cos(\theta)} \quad (8.5)$$

where d is the average particle size, λ is the wavelength, B is the full width at half maximum (FWHM) for a peak at 2θ (in radians), and θ is the corresponding Bragg's angle.

There are other more sophisticated methods to characterize the crystal structure of materials in a deeper way. For example, the Rietveld method uses powder

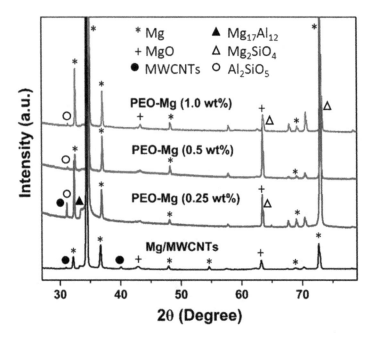

Fig. 8.7 XRD patterns of Mg-MWCNTs composites with a PEO coating

diffraction step-scanned intensities instead of integrated powder intensities [12]. This enables the full use of the information content of an XRD pattern. With this method, the structure determination and the quantitative phase analysis can be made with a high precision.

8.7 Raman Spectroscopy

Raman spectroscopy is a powerful and nondestructive technique used to give a structural fingerprint for the quantitative and qualitative identification of molecules. The technique has a wide range of applications in the analysis of materials, providing information on the molecular vibrations and crystal structures [13–15]. Bands observed in Raman spectroscopy are due to a change in the polarizability of the molecules produced by interaction with the light. Such an interaction produces specific vibrations that can be screened. Thus, for instance, the identification of single- and multiwalled carbon nanotubes and their diameter distribution and quality are possible to do with this technique.

In this same context, the discernment of different forms of carbon is here presented to show the Raman spectroscopy capabilities. The analysis was made by a micro Raman Horiba LAbRAM VIS-633 spectrometer provided with a He-Ne laser light (632.8 nm). Figure 8.8 shows the spectra of four carbon forms, which are used

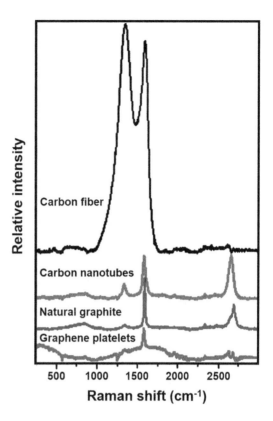

Fig. 8.8 Raman spectra of several carbonaceous structures

as reinforcing materials: carbon fibers, multiwalled carbon nanotubes, natural graphite, and graphene nanoplatelets. As can be seen, different bands are observed for each structure. Therefore, the technique allows to distinguish among each type of material, even though they are all made from carbon [13]. All spectra present the two main graphite bands, being much more visible for the MWCNTs. The G-band (1580 cm^{-1}) is characteristic of the in-plane vibration of the C-C bond. The D-band (1342 cm^{-1}) is activated by the presence of disorder in carbon systems. Here it is worth mentioning that the ratio of the D-band to G-band intensities (I_D/I_G) is useful to calculate the density of defects related to the crystalline quality in carbon samples [14]. Other bands can be also observed, such as the G´-band (2683 cm^{-1}), which corresponds to the overtone of the D-band [15]. A characteristic peak of the CNTs is that observed at 300 cm^{-1}, which corresponds to the radial breathing mode. This peak provides information about the electronic properties of the CNTs, and it is used for measuring their diameter, according to Eq. 8.6 [16].

$$\upsilon_{RBM} = \frac{248}{d_t} \tag{8.6}$$

where v_{RBM} is the radial breathing mode frequency and d_t is the diameter distribution of the sample.

Another important use for Raman spectroscopy is the capability of measuring the elastic modulus of CNTs in a matrix [17]. Although the development of the method concerns the use of an epoxy matrix, it can be used for other types of matrices. Young's modulus of carbon nanotubes may be derived from a concentric cylinder model for thermal stresses, using the D-band shift for each type of carbon nanotube (SWCNT or MWCNT).

8.8 Scanning Electron Microscopy and Energy-Dispersive Spectroscopy

Scanning electron microscopy (SEM) is a technique for the observation and characterization of both organic and inorganic materials from the micro (μm)- to the nano (nm)-scales. It is one of the most versatile techniques available for the examination and analysis of the microstructural features of solid objects, which takes advantage of the use of electrons to form high-resolution images with highly detailed information about the sample [18]. The re-emitted particles produced by the electron beam can be analyzed by different detectors, allowing the acquisition of information related to the morphology, surface defects, phase distribution, depth of field, and chemical composition. The preparation of SEM samples is relatively fast and easy, allowing the study of conductive and nonconductive samples by the application of a coating of carbon, gold, or gold/palladium, among others. There are scanning electron microscopes equipped with a field emission electron gun (FESEM), which have a better resolution, higher brightness, and reduced noise level than those of conventional ones. SEM and FESEM are useful for analyzing the phase distribution in light alloys and composites, the reinforcement distribution throughout the metallic matrix, and the matrix-reinforcement interfacial behavior.

Figure 8.9 shows some of the capabilities of FESEM. For example, images in Figs. 8.9a-b evidence the matrix morphology and how the MWCNTs are well embedded into the metallic matrix of both Al and Mg composites. The tensile fracture surface of a Mg-MWCNTs sample shows cleavage in Fig. 8.9c and how the MWCNTs act as bridges in the fractured matrix (Fig. 8.9d). Figure 8.9e displays the morphology of a porous $Ti_{50}Mg_{50}$ sample intentionally produced with high porosity. Finally, Fig. 8.9f presents a powder Ti6Al7Nb sample, showing the morphology and size of the particles.

Besides the morphological observations provided by SEM, a chemical characterization of samples is possible with the use of energy-dispersive spectroscopy (EDS) coupled to the scanning electron microscope [19, 20]. The EDS technique detects X-rays emitted from the material during the beam collision with its surface to determine the elemental composition of the analyzed volume. The screened X-ray energy is characteristic of the element from which it was emitted due to the fact that each

8.8 Scanning Electron Microscopy and Energy-Dispersive Spectroscopy

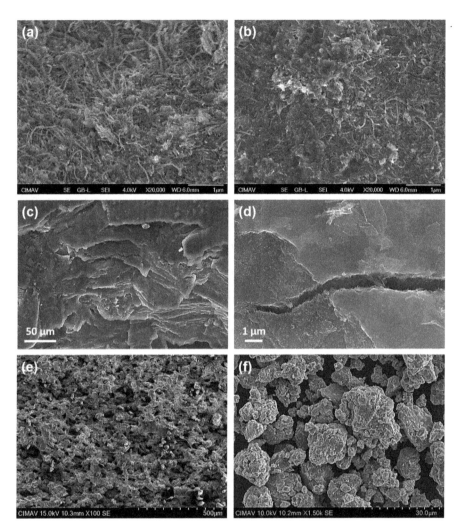

Fig. 8.9 Different samples analyzed by FESEM. (**a**) Al-MWCNTs and (**b**) Mg-MWCNTs composites produced by the sandwich technique. (**c**) Cleavage and (**d**) MWCNTs acting as bridge in a composite fractured in tensile test. (**e**) Porous $Ti_{50}Mg_{50}$ produced by powder metallurgy. (**f**) Ti6Al7Nb powders synthesized by mechanical alloying (*Image (e) courtesy: Dr. Caleb Carreño-Gallardo*)

element possesses a unique atomic structure, obtaining a unique set of peaks on its X-ray spectrum. This technique helps to identify the elemental composition of the phases formed during the process. The technique uses an ultrathin window detector, and elements up to boron can be detected with a very good sensitivity.

Figure 8.10 presents the SEM-EDS results of the powder Ti6Al7Nb alloy processed by mechanical alloying. The spectrum presents the expected elements (Ti,

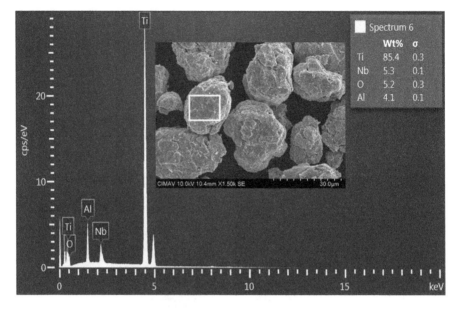

Fig. 8.10 SEM-EDS characterization of the mechanically alloyed Ti6Al7Nb alloy

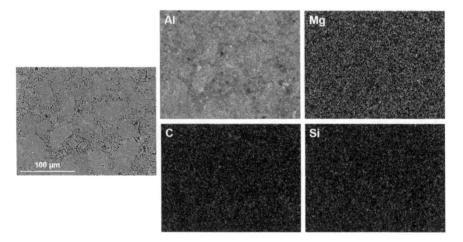

Fig. 8.11 SEM-EDS elemental mapping analysis of an Al6061-fullerene soot composite

Al, and Nb), along with oxygen as a contamination, which can come from the processing and/or the environment.

The EDS technique is capable of producing elemental distribution maps. As an example, Fig. 8.11 presents the SEM-EDS elemental mapping of an Al6061 alloy reinforced with fullerene soot. This mapping allows to see how the alloying elements are distributed along the aluminum matrix. Carbon comes from the fullerene soot, which is homogeneously distributed through the matrix.

8.9 TEM Sample Preparation

In comparison with SEM, the acquisition of images by TEM is performed at a resolution as high as atomic level. However, the preparation of high-quality specimens is more elaborated in TEM. Indeed, techniques such as electropolishing, ultramicrotomy, and focused ion beam are used to obtain ultrathin sample films [21, 22]. In addition, the use of metal grids is required to support the samples.

8.9.1 Electropolishing

Electropolishing, also known as electrolytic polishing, is a technique by which the smoothing of a metallic surface is achieved. For the preparation of TEM samples by this technique, a slice of the sample is cut as thin as possible by a diamond disk or similar. The sample slice is fastened to a device and then polished with 1500–4000 grit (depending on the material) SiC emery paper to reduce its thickness to 100–300 μm. The sample is removed from the device, cleaned, and dried. A 3-mm-diameter foil is created using punch tools for ductile samples or ultrasonic cutter for brittle samples. This foil is placed into the electropolishing equipment holder, where it will be thinned until perforation by a chemical reagent jet released by a nozzle. For this task, the equipment is set up with a voltage range to provide the necessary current density, which depends on the type of material to be electropolished and the chemical reagent used. Figure 8.12a presents a photography of an electropolishing equipment, Fig. 8.12b represents the holder-sample-nozzle setup, and Fig. 8.12c shows an electropolished aluminum sample.

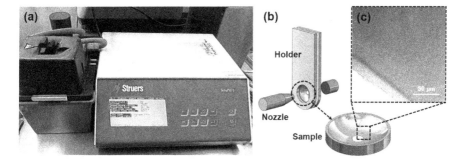

Fig. 8.12 (a) Electropolishing equipment for the sample preparation. (b) Scheme of the holder-sample-nozzle setup. (c) Electropolished Al sample (*Image (c) courtesy: Dr. Hansel Manuel Medrano-Prieto*)

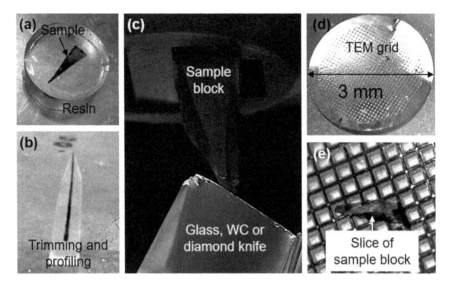

Fig. 8.13 Procedure to the sample preparation by ultramicrotomy

8.9.2 Ultramicrotomy

The ultramicrotomy technique consists in cutting samples into ultrathin sections. As an example of the application of the ultramicrotomy, a RMC Boeckeler PT-PC Power Tome ultramicrotome provided with a diamond knife was used to obtain films of ~100 nm in thickness. A sample is taken from the specimen to be analyzed, which is mounted in resin to form a sample block (Fig. 8.13a). The sample block is trimmed and profiled for having a ~ 1-mm block face (Fig. 8.13b). The sample block is looked at the sample holder (Fig. 8.13c). The block can be observed by using a binocular stereomicroscope coupled to the ultramicrotome. A fine profile cut is performed setting an angle on the knife and making a few cuts; the knife can be made from glass, WC or diamond. After this, the desired sections are cut and received in a container full of water or alcohol. Finally, the sections are mounted on a TEM grid (Figs. 8.13d-e) and air-dried.

8.9.3 Focused Ion Beam

The focused ion beam (FIB) technique [21] is very useful for the preparation of TEM samples. The technique works by accelerating thermal ions extracted from a Ga filament. The ions impact on a selected area, with a determined energy and scanning speed. After a certain number of passes, the work can be inspected using secondary electron images, similar to those generated by SEM. FIB allows to select

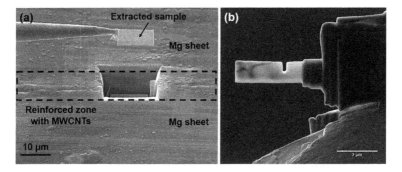

Fig. 8.14 Mg-MWCNTs composite. (**a**) Zone where the sample was extracted. (**b**) Sample in its final state of preparation

preferred positions in the samples; therefore specific mechanical properties can be evaluated [23]. The advantages of using a FIB for TEM specimen preparation are multiple [24]:

- There is no other technique with which a precise selection of the target area can be reached, i.e., lamellae can be placed with an accuracy of ~20 nm.
- The preparation is fast and reliable, i.e., in just ~20 min, and within a maximum of 2–4 h, specimens of a wide range of materials can be prepared.
- FIB preparation techniques are independent of the material nature.

As an example of the application of the FIB technique, a focused ion beam (FIB Carl Zeiss, Germany) system was used to obtain micro-cantilever beams of Mg-MWCNTs composites. A high current Ga^+-ion beam (30 keV, 7 nA, 700 and 300 pA), followed by a fine milling at low currents (30 keV, 50 pA) were used. Figure 8.14 presents the preparation of a sample by FIB, showing the zone where it was extracted and its final form.

8.10 Transmission Electron Microscopy

The transmission electron microscopy (TEM) technique uses transmitted electrons, i.e., electrons that pass through the sample, to generate an image. The sample must have an ultrathin thickness of ~100 nm for an appropriate analysis. The interaction of the electron beam with the material produces diffraction of the transmitted electrons in a coherent way to the crystalline planes of the material. Thus, it is possible to obtain a diffraction pattern of ordered points with information concerning the orientation and crystalline structure of the sample [25].

An example of the characterization by TEM is shown in Fig. 8.15 for a $Ti_{75}Mg_{25}$ alloy synthesized by mechanical alloying. The presence of nanograins (below 20 nm) can be seen, and the selected area electron diffraction (SAED) pattern gives

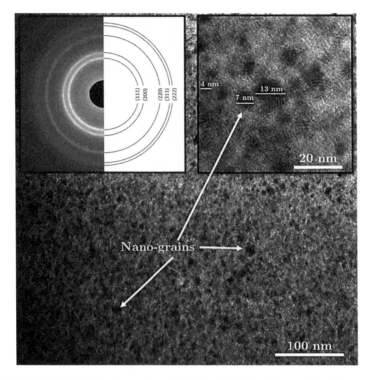

Fig. 8.15 Bright field TEM micrograph and SAED of $Ti_{75}Mg_{25}$ alloy produced by mechanical alloying

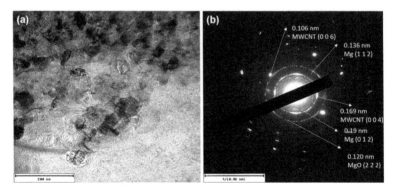

Fig. 8.16 (a) TEM bright field image and (b) SAED pattern of a Mg-MWCNTs composite

information concerning the crystal structure of the material; in this case, the sample has an fcc structure.

Another example is shown in Fig. 8.16 for a Mg-MWCNTs composite. The TEM bright field image (Fig. 8.16a) shows the presence of different grain sizes and MWCNTs well embedded in the metal matrix. The SAED pattern (Fig. 8.16b) reveals the presence of Mg, MWCNTs, and MgO; the latter is seen as small grains in Fig. 8.16a.

8.11 High-Resolution Transmission Electron Microscopy

The high-resolution transmission electron microscopy (HRTEM) technique provides direct images of the atomic structure of samples [26], giving crystallographic information of the materials. The technique is widely used in advanced characterization of materials, allowing the acquisition of information about punctual defects, stacking faults, precipitates, and grain boundaries.

Figures 8.17 and 8.18 show HRTEM images of Al-MWCNTs and Mg-MWCNTs composites, where nanograins and MWCNTs well embedded in the metallic matrix can be seen. There is no evidence of the phase formation at the interface between the matrix and MWCNTs. In addition, some other aspects can be appreciated in

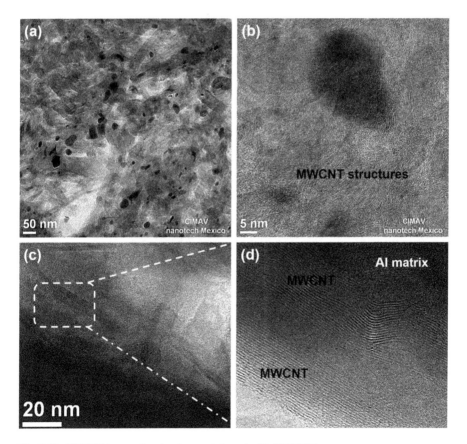

Fig. 8.17 HRTEM images for aluminum reinforced with MWCNTs

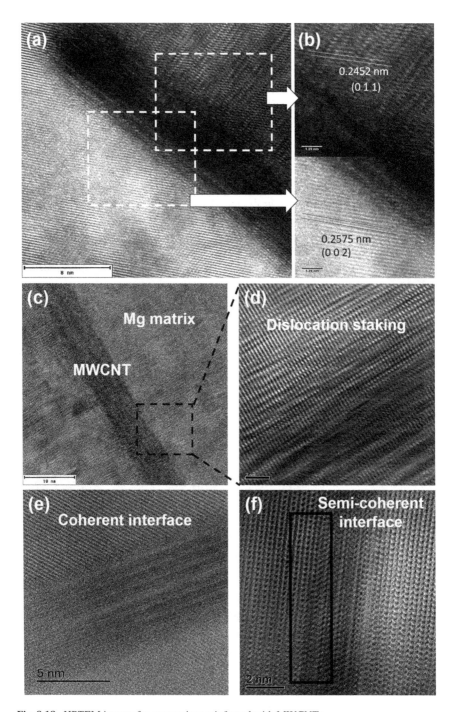

Fig. 8.18 HRTEM images for magnesium reinforced with MWCNTs

8.12 Electron Energy Loss Spectroscopy (EELS)

Fig. 8.18, such as interplanar distances, the crystalline orientation, stacking faults, dislocation formation and stacking, coherent and semi-coherent interfaces, and interlayer spacing of the MWCNTs. HRTEM images can reveal the presence of different strengthening mechanisms.

8.12 Electron Energy Loss Spectroscopy (EELS)

The electron energy loss spectroscopy (EELS) technique is based on the inelastic scattering of fast electrons impinging in a thin specimen [27]. The electron beam loses energy and is bent through a small angle (5–100 mrad). The energy distribution of all the inelastically scattered electrons provides information about the local environment of the electrons, which in turn relates to the physical and chemical properties of the material. EELS provides structural and chemical information about a material at high resolution, being able to detect low atomic number elements and provide detailed information about the electronic state and chemical bonding of a sample.

Elemental analysis by EELS can be performed, for instance, in a long line as small as the diameter of CNTs and carbon onions. EELS can provide information on the transformation of the CNT surface when they are in contact with other material, such as metallic matrices. An example of the technique application is to identify the semi-coherent interface in Fe-filled CNTs [28]. Figure 8.19 shows the EELS spectrum and elemental mapping of magnesium reinforced with MWCNTs. The results suggest the presence of aluminum close to the MWCNTs walls. Aluminum has chemical affinity for carbon, which may result in the formation of aluminum carbide, but no interphase was found in this case.

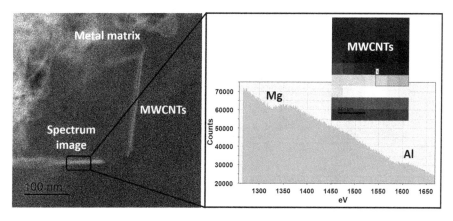

Fig. 8.19 (**a**) TEM image of a Mg-MWCNTs composite and (**b**) EELS spectrum and elemental mapping at the metallic matrix-MWCNTs interface

8.13 X-Ray Photoelectron Spectroscopy

X-ray photoelectron spectroscopy (XPS), also known as electron spectroscopy for chemical analysis (ESCA), is a technique used to analyze the surface of materials [29]. The XPS technique is classified within the analytical techniques of electronic spectroscopy because electrons are measured. It provides information on the elemental composition, chemical and electronic states, and empirical formula of the elements within the material surface. When an X-ray beam irradiates a material, a spectrum is generated, and the kinetic energy of electrons emitted from the surface (1–10 nm) is measured. The horizontal axis of the XPS spectrum shows binding energy values, and the vertical axis represents the intensity or measured counts. The peaks correspond to electrons of a particular characteristic energy; this energy, along with the peak intensity, enables the identification and quantification of the surface elements.

An example of the XPS technique application is presented here for Ti-Mg alloys synthesized by mechanical alloying and sintering. Samples were polished with SiC emery paper and subsequently polished using a cloth with no other substance. Ethanol was used as a lubricant because milled Ti and Ti-Mg compositions react with water. After polishing, the samples were cleaned with ethanol and dried in a furnace at 80 °C for 8 h. Then they were placed under vacuum to evacuate the encapsulated air from the pores and to avoid further contamination. To remove the undesirable elements from the sample surface, an Ar sputtering on the samples was performed before the XPS analysis.

Figure 8.20a shows the XPS spectrum of the $Ti_{75}Mg_{25}$ alloy. Figure 8.20b shows the deconvolution of the Ti2p peak, where TiC and titanium oxides are detected. The main peak of Mg1s (1304.5 eV) in Fig. 8.20c reveals the formation of magnesium oxide, with no signal of metallic magnesium (1303 eV). However, Mg1s peak is usually accompanied by Auger signals in the range of

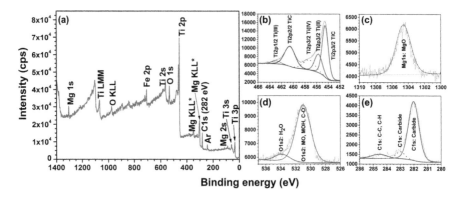

Fig. 8.20 (a) XPS spectrum of the $Ti_{75}Mg_{25}$ alloy. Peak deconvolution of (b) Ti2p, (c) Mg1s, (d) O1s and (e) C1s

300–390 eV. When Mg is buried under C, Mg KLL Auger peaks will be observed even if Mg1s is not present; this behavior is due to a difference in electron kinetic energy. The main signal of Mg KLL Auger peak (300–306 eV) has large chemical shifts and may be useful for the chemical state analysis. According to this information, the binding energies coming from the Mg Auger electrons match with Mg in its metallic state. The presence of oxygen and carbon is noticed in Fig. 8.20a. The oxygen is part of the atmosphere, and it is naturally deposited on the surface of the samples. The O1s peak confirms the formation of a metallic oxide (Fig. 8.20d). The powder processing by mechanical alloying may generate some metal oxides like TiO_2 and MgO. Carbon can come from the residual stearic acid used as a process control agent during the mechanical alloy process. The presence of TiC (Fig. 8.20b) is confirmed in Fig. 8.20e, where the C1s peak matches with the formation of this carbide; C-C and C-H bonds are also present.

8.14 Mechanical Properties

The mechanical properties of a material are those properties that involve a reaction to an applied load. Different techniques are used for the mechanical characterization of materials. Among them, tensile, compression, bending, impact, fatigue, creep, and hardness tests are the most common ways to evaluate the mechanical behavior of materials. The mechanical properties of a material are not constants and often change as a function of temperature, rate of loading, and other external conditions. It should also be noted that there is often significant variability in the values obtained when measuring mechanical properties. A seemingly identical test specimen from the same lot of material will often produce considerably different results. Therefore, multiple tests are commonly conducted to determine the mechanical properties, and the values reported are the average value or calculated statistical minimum value.

The mechanical properties can be evaluated at different scales, going from the bulk scale, through the microscale and to the nanoscale. Most of the bulk-scale tests are well established and standardized. AFNOR, AISI, ANSI, AS, ASME, ASTM, BS, DIN, and JIS are the most known standard institutes and organizations worldwide. However, the development of micro- and nanostructures requires new tools for their mechanical characterization. Different testing techniques have been widely used to evaluate the mechanical properties of materials at the small scale, being of interest both in academic and engineering applications. The mechanical testing challenges at the small scale derive from the very basic experimental issues, such as the specimen preparation and manipulation, and high-resolution load and displacement sensing, to complex experimental issues, such as enhanced multiphysics coupling and the specimen-environment interaction [30].

8.14.1 Bulk-Scale Mechanical Testing

The mechanical properties of bulk light alloys and composites are of great interest for their application in structural applications in several industries, including the transportation industry. As an exercise of the bulk mechanical characterization, tension and tribological tests of light metal matrix composites will be presented below. In this kind of composites, the mechanical behavior depends on different factors inherent to the reinforcing material, such as its dispersion and aligning.

8.14.1.1 Tensile-Compression Tests

The evaluation of the mechanical behavior of a sample under conditions of tension and compression can be performed to provide basic material property data, which are critical for the component design and service performance assessment. The requirements for tensile and compressive strength values and the methods for measuring these properties are specified in various standards for a wide variety of materials. Testing can be performed on machined material samples or on full-size or scale models of actual components. These tests are typically performed using universal testing machines.

A tensile test is a method for determining the behavior of materials under axial tensile loading. The test is conducted by fixing the specimen into the test apparatus and then applying it a load by separating the testing machine crossheads. A compression test is a method for determining the behavior of materials under a compressive load. In this test, the specimen experiences an opposite load that push inward upon it from opposite sides. The crosshead speed in both tests can be varied to control the rate of strain in the test specimen. Data from the tests are used to determine tensile/compression strength, yield strength, and modulus of elasticity.

In this case study, tensile tests were performed for aluminum and magnesium composites fabricated by the sandwich technique. Figure 8.21 shows that the addition of MWCNTs significantly improves the yield strength, ultimate tensile strength, and elastic modulus for both groups of composites, with respect to the matrix with no reinforcement. This increment in properties indicates that there is a good dispersion of the MWCNTs and a good interfacial bonding among them and the metallic matrix, which generates a good load transfer from the matrix to the MWCNTs.

8.14.1.2 Tribological Tests

The tribological evaluation is an important tool to study the frictional and wear behavior in light alloys and composites. For composites reinforced with carbonaceous materials, which act as lubricants in some matrices, the tribological tests play an important role to determine their effectiveness. CNTs not only strengthen the metallic matrix but also reduce the coefficient of friction. Tribological properties of

8.14 Mechanical Properties

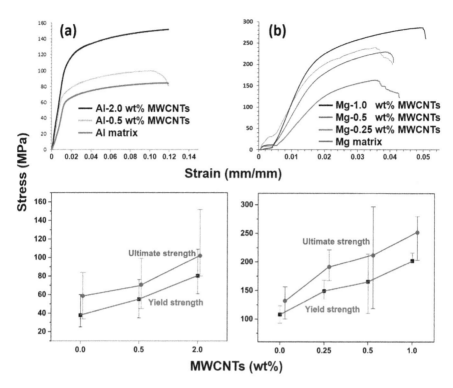

Fig. 8.21 Tensile mechanical properties for (**a**) aluminum and (**b**) magnesium composites

metal matrix composites are evaluated using conventional techniques like ball-on-disk [31, 32], ring-on-block [33, 34], and pin-on-disk [35–40]. The latter was used in this chapter to characterize the AZ31B magnesium alloy reinforced with MWCNTs. The tribological tests were performed according to the ASTM G99 standard [41] in a pin-on-disk test machine. For that, rectangular-section pins of AZ31B-MWCNTs composites and flat disks of the AZ31B alloy were used. Figure 8.22a schematically shows the tribological test setup. The samples for the test were previously polished with 1-μm alumina and ultrasonically washed. A load of 5 N, a linear velocity of 0.26 m/s, and a time of 5 min were set for the tests. The micrograph in Fig. 8.22b shows the appearance of the sample after the test.

Figure 8.23 shows the results of the tribological and nanoindentation tests with respect to the MWCNTs content. The wear rate graph of the AZ31B-MWCNTs pins (Fig. 8.23a) does not show a significantly different pattern among the AZ31B matrix and the AZ31B-MWCNTs composites. However, the wear rate of the unreinforced AZ31B disks (Fig. 8.23b) increases as the MWCNTs content in the pin (AZ31B-MWCNTs) increases. This behavior is because the pin has better mechanical properties than those of the unreinforced disk, which promotes the plastic deformation on the disk, as shown in Figs. 8.23c-d.

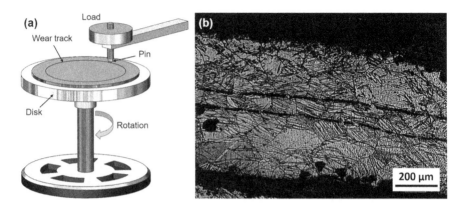

Fig. 8.22 (a) Scheme of the tribological test setup. (b) Evaluated zone of the AZ31B-MWCNTs composite

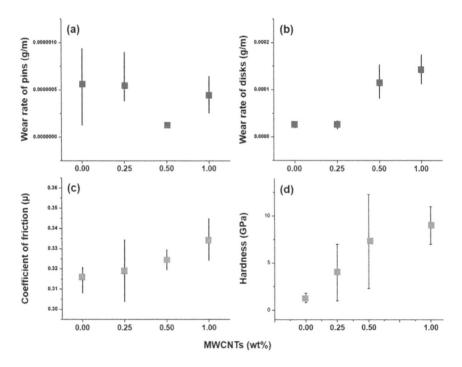

Fig. 8.23 Tribological and mechanical behavior of AZ31B-MWCNTs composites. (a) Wear rate of pins. (b) Wear rate of the disks. (c) Coefficient of friction of the system. (d) Hardness of the pins

8.14.2 Nano- and Micromechanical Testing

Nano- and micromechanical testing of light alloys and composites can be performed by different and novel characterization techniques, including nanoindentation, nanoscratch, atomic force microscopy, and micro-cantilever tests. The development of new transducers allows to have the appropriate resolution for measuring the mechanical properties in small areas of samples.

8.14.2.1 Nanoindentation

The nanoindentation technique has been widely used to characterize the mechanical properties of materials. It has advantages in probing the properties at the submicron scale. Nanoindentation is able to measure mechanical properties like hardness and elastic modulus of samples. It does not need a complicated sample preparation and can measure properties of various materials, ranging from hard superalloys to soft biomaterials in seconds, making it the fastest technique for such measurements. The technique has a great potential in the mechanical characterization of nanoreinforced metal matrices; several studies can be performed since the response is generated from submicrometric size regions.

As a case study, nanoindentation tests were performed to analyze the mechanical behavior of AZ31B-MWCNTs composites synthesized by the sandwich technique. Figure 8.24a shows the indentations made near the interface between magnesium sheets. All tests were made with the same load (1 mN); therefore no indentation size effect due to this factor is involved. Figure 8.24b presents the resulting load-displacement (P-h) curves for composites reinforced with 0, 0.25, 0.5, and 1.0 wt% MWCNTs, which were used to calculate the elastic modulus and hardness, according to the Oliver and Pharr method [42].

Figures 8.24c-d show the results of the elastic modulus and hardness, respectively, with respect to the MWCNTs content. It is worth mentioning that such values for the AZ31B matrix (0 wt% MWCNTs) correspond to those of the bulk AZ31B reported in the literature; consequently, at this load level, there is no effect at all. As can be seen, both the elastic modulus and hardness were significantly increased close to the interface for all the MWCNTs contents. Compared with the magnesium sheets stacked with no MWCNTs, the increase in elastic modulus was 85%, 100%, and 115% and in hardness was 220%, 420%, and 600% for the composites reinforced with 0.25, 0.5 and 1.0 wt% MWCNTs, respectively. This indicates that the reinforcement is responsible for the improved mechanical properties. Figures 8.24e-f show the variation of the elastic modulus and hardness as a function of the distance in the diffusion zone.

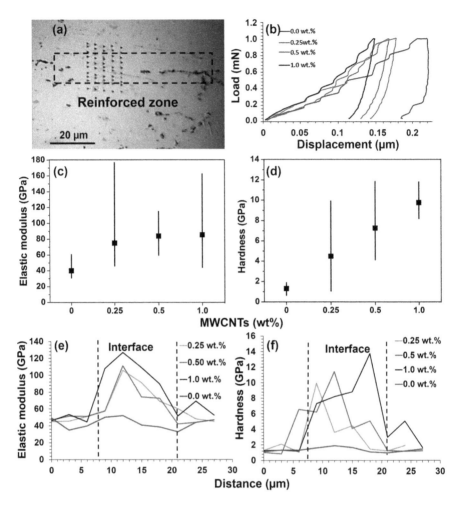

Fig. 8.24 (a) Nanoindentations performed close to the interface between Mg sheets. (b) P-h curves. (c–e) Elastic modulus. (d–f) Hardness

8.14.2.2 Modulus Mapping

Modulus mapping is a plot of the spatial variation of the elastic modulus for a composite, in order to identify the reinforcement distribution or mechanical property homogeneity. One of the several ways to make a modulus mapping is by measuring the mechanical properties while the nanoindenter tip is rastering the surface. Then a scanning probe image is generated during the application of the dynamic force. The phase and amplitude of the dynamic signal is measured by the lock-in amplifier and used to calculate the storage and loss modulus at the given pixel by Eqs. 8.7 and 8.8.

8.14 Mechanical Properties

$$X_0 = \frac{F_0}{\sqrt{(k - mw^2) + \left[(C_i - C_s)w\right]^2}} \quad (8.7)$$

$$\emptyset = \tan^{-1}\frac{(C_i + C_s)w}{k - mw^2} \quad (8.8)$$

where X_0 and $\emptyset$ refer to the amplitude and the phase difference between the force and displacement, F_0 is the force amplitude, w is the frequency, m is the indenter mass, C_i is the coefficient of the air gap between the plates and capacitive transducer, C_s is the damping coefficient of the sample, and k is the combined stiffness given by $K_i + K_s$, where K_i is the spring constant of the spring holding the indenter and K_s is the contact stiffness. The elastic modulus is related to the contact stiffness by Eq. 8.9.

$$K_s = 2E^* h \sqrt{\frac{24.5}{\pi}} \quad (8.9)$$

where E^* is the reduced modulus and h is the penetration depth.

This force modulation technique provides the elastic modulus at each image pixel resulting in 65,536 values per image, which can be used to prepare the modulus mapping over the selected microstructure, as was studied by Laha et al. [43] in Al-Si nanocomposites reinforced with CNTs.

Another way of making a modulus mapping is by several standard indentations with high resolution, i.e., low loads and minimum displacements of the tip [44]. The nanoindentation tests were performed on an IBIS Authority Fischer-Cripps with a Berkovich-type diamond tip in closed loop mode. Two areas were mapped. First area was of 50 μm × 25 μm with a maximum load of 1 mN, and the second area was of 5 μm × 5 μm with a maximum load of 0.3 mN. The mechanical properties were deconvoluted using the Oliver and Pharr method [42, 45, 46] and then plotted in contour curves, as shown in Fig. 8.25. These results evidence the modulus distribution into the Mg matrix fabricated by the sandwich technique. The elastic modulus ranges from ~45 GPa for the base material to ~120 GPa for the reinforced zone for all composites studied. In addition, the images show that the reinforcing phase is properly dispersed into the metallic matrix.

8.14.2.3 Nanoscratch

Nanoscratch tests can provide information about the tribological behavior such as wear resistance and coefficient of friction in materials and composites at the nanoscale level. The technique involves scratching the sample surface using a diamond tip indenter at low loads (a few mN).

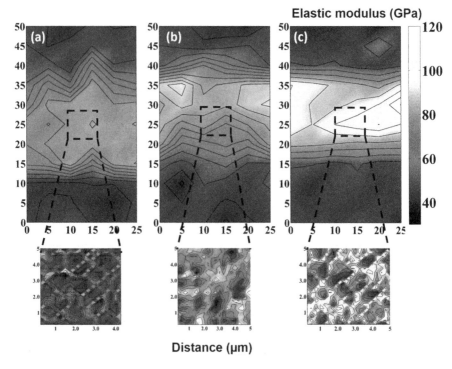

Fig. 8.25 Nanoindentation mappings in Mg reinforced with (**a**) 0.25, (**b**) 0.5, and (**c**) 1.0 wt% MWCNTs

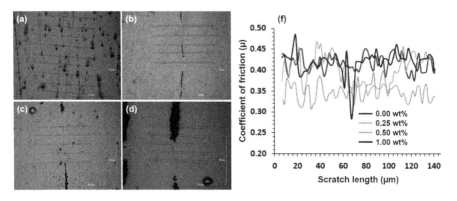

Fig. 8.26 Nanoscratch tests for Mg composites containing (**a**) 0.0, (**b**) 0.25, (**c**) 0.5, and (**d**) 1.0 wt% MWCNTs. (f) Variation of the coefficient of friction as a function of the scratch length

Figures 8.26a–d show optical microscopy images of the tested samples, where the nanogroove product of the tests can be observed. Figure 8.26e presents the variation of the coefficient of friction as a function of the scratch length. For all the composites studied, the coefficient of friction decreased in the study zone. The decrease was from 0.45 for Mg with no reinforcement to 0.35, 0.33, and 0.28 for composites reinforced with 0.25, 0.5, and 1.0 wt% MWCNTs, respectively.

8.14.2.4 AFM Mechanical Testing

Atomic force microscopy (AFM), also known as scanning force microscopy (SFM), is a very high-resolution technique used for obtaining quantitative and qualitative data based on different properties like morphology, size, surface roughness, and texture. AFM utilizes a micro-cantilever as a scanning probe. A sensor scanner is used to create precision nanoscale displacements of the sample and cantilever. A laser beam is employed to monitor the cantilever movement when it is in relative motion between materials. A laser light is reflected off the flexible, free end of the cantilever, onto a position-sensitive photodetector. The feedback system is used to maintain the cantilever at vertical position, permitting the acquisition of surface topography during scanning.

As an example of the capabilities of the AFM technique, the mechanical properties of the Al7075 alloy reinforced with the addition of graphene (GNp), functionalized graphene (f-G), carbon nanotubes (CNT), and functionalized carbon nanotubes (f-CNT) were evaluated. The composites were synthesized by mechanical alloying for 5 h and 20 h and sintered at 550 °C for 3 h. The nanoindentations were performed with a Hysitron model T1950 triboindenter equipment using a load of 2000 μN and coupled to a high-resolution scanning probe microscope for accurate observations of the deformations after testing; 12 nanoindentations were performed on each sample in a 15 μm × 15 μm field with a Berkovich type diamond tip. The Vickers microhardness of samples was evaluated with a Matsuzawa MMT-X7 equipment using a load of 1 kg and a symmetrical diamond-shaped indenter with a square base and 136° on the faces; 10 indentations were performed on each sample, and the indentation size was measured using an optical microscope according to ASTM E384 [47].

Figure 8.27 shows the results of the Vickers microhardness of the synthesized nanocomposites. As can be seen, the milling time increases the hardness up to 75% (5 h) and 20% (20 h) for the Al7075–1 wt% f-CNT nanocomposite, with respect to the Al7075 alloy. Therefore, a longer milling time produces not only the reduction in the particle size along with severe deformations, but also these deformations generate a great amount of dislocations in the crystalline network, increasing the mechanical strength and hardness of the materials.

Nanoindentation tests were achieved in samples synthesized for 20 h of milling. The results are presented in Fig. 8.28, which confirm the superiority in properties of the Al7075–1 wt% f-CNT sample. The Young's modulus and hardness were increased by 40.1% and 89.6%, respectively, compared to the Al7075 alloy. The increment in mechanical properties was by the incorporation of functionalized carbon nanotubes, which were found homogeneously dispersed in the metal matrix.

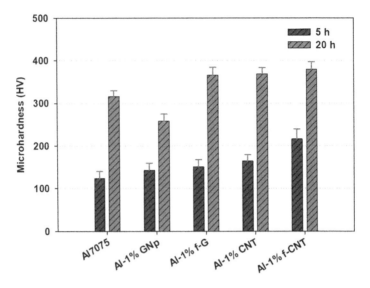

Fig. 8.27 Vickers microhardness of different Al7075 samples

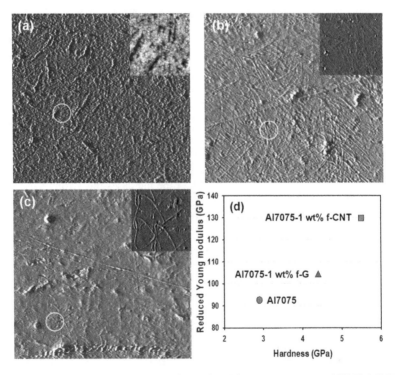

Fig. 8.28 AFM images of (**a**) Al7075 alloy, (**b**) Al7075-f-G nanocomposite, (**c**) Al7075-f-CNT. (**d**) Comparison of reduced Young's modulus and hardness

8.14.2.5 In Situ Micro-Cantilever Beam Test

The cantilever beam test is used to measure the bending strength and modulus of elasticity in bending of a specimen. To perform the test, one end of the beam is fixed, and the other end is free, on which a load is applied until failure occurs (Fig. 8.29). The test method ASTM E399 [48] covers the determination of fracture toughness (K_{IC}) of metallic materials by tests using a variety of specimens. Among these, the notched specimens are considered.

For the study of the mechanical interaction between matrix and reinforcement at the small scale, micro- and nanomechanical test methods are required, which pose a challenge. While some mechanical testing methods are available, vast efforts are still undertaken to improve them and develop new methods for reliable mechanical measurements at such scale.

In this section, an exercise using microscale cantilever beam specimens is presented. AZ31B magnesium reinforced with 0.25, 0.5 and 1.0 wt% MWCNTs were prepared by focused ion beam (FIB), as explained in Sect. 8.9.3. Figure 8.30a presents the dimensions of the notched micro-cantilever beam. For the micro-cantilever beam test, a Nanofactory Instruments nano-force AFM sensor intended for in situ TEM measurement was used. Figures 8.30b-c show the setup of the test, before and after testing the sample. The in situ microscale testing presents the advantage of being able to observe the response of the specimen during the test.

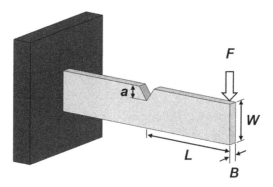

Fig. 8.29 Schematic of a notched cantilever beam

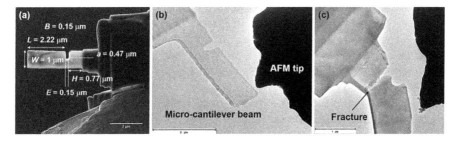

Fig. 8.30 Micro-cantilever beam test: (**a**) sample dimensions, (**b**) beginning, and (**c**) end of the test

The fracture toughness K_{IC} of samples is calculated using Eq. 8.10.

$$K_{IC} = \frac{F_{max} L}{BW^{\frac{3}{2}}} F\left(\frac{a}{W}\right) \tag{8.10}$$

where F_{max} is the fracture force and $F(a/W)$ is a dimensionless geometry factor of the tested specimens [48]. The loading span L is the distance between the notch and the loading point, a is the crack length until prenotch plus initial crack length, W is the cantilever width, and B is the thickness of sample. The geometry factor $F(a/W)$ may be used for a/w ratios between 0.05 and 0.55, according to Eq. 8.11.

$$F\left(\frac{a}{W}\right) = 4 \frac{3\left(\frac{a}{w}\right)^{0.5}\left[1.23 - \left(\frac{a}{w}\right)\left(1 - \left(\frac{a}{w}\right)\right)\left(-6.09 + 13.96\left(\frac{a}{w}\right) - 14.05\left(\frac{a}{w}\right)^{2}\right)\right]}{2\left(1 + 2\left(\frac{a}{w}\right)\right)\left(1 - \left(\frac{a}{w}\right)\right)^{1.5}} \tag{8.11}$$

The calculated value of the fracture toughness was 17.21, 21.04, and 20.13 MPa√ŭ5A; for the composites reinforced with 0.25, 0.5, and 1.0 wt% MWCNTs, respectively. These results suggest a dependency of the fracture toughness with the MWCNTs content, which is due to the different strengthening mechanisms involved.

8.15 Conclusions

In this chapter, different characterization techniques have been described and discussed in a manner to show the reader the capability and limitations while characterizing light alloys and composites. These characterization techniques were divided into two types: (1) chemical, structural, and microstructural characterization and (2) mechanical characterization.

An important technique for the accurate chemical analysis is the inductively coupled plasma optical emission spectrometry. Physical properties, such as thermal and density properties, are important for identifying the interactions, for instance, between the matrix and the reinforcement in a composite; one of the most accessible and economical techniques is the Archimedes' method.

The structure of light alloys and composites can be characterized by X-ray diffraction and Raman spectroscopy. XRD can detect the phases present in light alloys and composites, certain interactions between the metallic matrix and the reinforcement, the fiber orientation in a composite, the effect of the reinforcement on the grain size, and the dislocation density. Raman spectroscopy is used as the most conclusive evidence on the presence of a reinforcement in a specific matrix, such as that of CNTs and their morphological characteristics.

The microstructural behavior can be described at different scales by optical microscopy, scanning electron microscopy, transmission electron microscopy, and high-resolution transmission electron microscopy. Both the SEM and TEM techniques have the capability of making elemental analysis by energy-dispersive spectroscopy. TEM and HRTEM can provide information on metal matrix composites, such as the wettability, interphase formation, and orientation relationship between the metallic matrix and the reinforcement. This is a very useful information for the elucidation of the strengthening mechanisms in metal matrix composites. Additionally, with high-magnification HRTEM is possible to use the electron energy loss spectroscopy to understand the possible reactions between the metallic matrix and reinforcement; this characterization can also be done by X-ray photoelectron spectroscopy.

Mechanical characterization of light alloys and composites provides information on the strength, elastic modulus, and fracture toughness, among other properties. Several techniques are used for performing this characterization at different scales: tensile, compression, macrohardness, microhardness, nanoindentation, and AFM tests, among others. The macroscale tests give bulk information about the materials, with no discrimination between the matrix and the reinforcement in the case of composites. The micro- and nanoscale tests can identify and discriminate the features of the matrix and reinforcement. In situ characterization techniques are used to understand the strengthening or fracture mechanisms in light alloys and composites, as well as to quantify the micro- and nanomechanical properties.

In summary, conventional and advanced characterization techniques are necessary for understanding the chemical, structural, microstructural, and mechanical behaviors of light alloys and composites. Advanced techniques are especially required when the reinforcing materials have micron, submicron, and nanodimensions.

References

1. Armstrong, R., Adams, B. L., & Arnold, M. A. (2018). *ASM handbook volume 10, materials characterization*. Materials Park: ASM International.
2. Cayless, R.B.C. (1990). *ASM handbook volume 2, properties and selection: Nonferrous alloys and special purpose materials*. Materials Park, OH: ASM International
3. Hernandez Robles, F. C., Ramirez, J. M. H., & Mackay, R. (2017). *Al-Si alloys: Automotive, aeronautical, and aerospace applications*. Cham: Springer.
4. Spierings, A. B., Schneider, M., & Eggenberger, R. (2011). Comparison of density measurement techniques for additive manufactured metallic parts. *Rapid Prototyping Journal, 17*(5), 380–386.
5. Complementing X-ray *imaging with Neutron Radiography*. 2020.; Available from: https://andor.oxinst.com/learning/view/article/complementing-x-ray-imaging-with-neutron-radiography.
6. *Sartorius YDK 01, YDK 01-0D, YDK 01 LP Density Determination Kit User's Manual*. 1998. Sartorius AG: Goettingen, Germany.

7. Latief, F., & Sherif, E.-S. M. (2012). Effects of sintering temperature and graphite addition on the mechanical properties of aluminum. *Journal of Industrial and Engineering Chemistry, 18*(6), 2129–2134.
8. Woods, S. (2015). 3-D CT inspection offers a full view of microparts. Available from: http://micromanufacturing.com/showthread.php?t=876.
9. Du Plessis, A., et al. (2018). Standard method for microCT-based additive manufacturing quality control 2: Density measurement. *MethodsX, 5*, 1117–1123.
10. Guinebretière, R. (2013). *X-ray diffraction by polycrystalline materials*. Somerset: Wiley.
11. Cao, A., et al. (2001). X-ray diffraction characterization on the alignment degree of carbon nanotubes. *Chemical Physics Letters, 344*(1–2), 13–17.
12. Rietveld, H. M. (2014). The Rietveld method. *Physica Scripta, 89*(9), 098002.
13. Dresselhaus, M. S., et al. (2010). Perspectives on carbon nanotubes and graphene raman spectroscopy. *Nano Letters, 10*(3), 751–758.
14. Dresselhaus, M. S., et al. (2005). Raman spectroscopy of carbon nanotubes. *Physics Reports, 409*(2), 47–99.
15. Heise, H. M., et al. (2011). Characterization of carbon nanotube filters and other carbonaceous materials by Raman spectroscopy—II: Study on dispersion and disorder parameters. *Journal of Raman Spectroscopy, 42*(3), 294–302.
16. Jorio, A., et al. (2001). Structural (n, m) determination of isolated single-wall carbon nanotubes by resonant Raman scattering. *Physical Review Letters, 86*(6), 1118–1121.
17. Lourie, O., & Wagner, H. (1998). Evaluation of Young's modulus of carbon nanotubes by micro-Raman spectroscopy. *Journal of Materials Research, 13*(9), 2418–2422.
18. Goldstein, J. I., et al. (2017). *Scanning electron microscopy and X-ray microanalysis*. New York: Springer.
19. Girão, A. V., Caputo, G., & Ferro, M. C. (2017). Application of scanning electron microscopy–energy dispersive X-ray spectroscopy (SEM-EDS). *Comprehensive Analytical Chemistry, 75*, 153–168.
20. Abd Mutalib, M., et al. (2017). Chapter 9 – Scanning electron microscopy (SEM) and energy-dispersive X-ray (EDX) spectroscopy. In *Membrane characterization* (pp. 161–179). Amsterdam: Elsevier.
21. Lotnyk, A., et al. (2015). Focused high- and low-energy ion milling for TEM specimen preparation. *Microelectronics Reliability, 55*(9–10), 2119–2125.
22. Ünlü, N. (2008). Preparation of high quality Al TEM specimens via a double-jet electropolishing technique. *Materials Characterization, 59*(5), 547–553.
23. Darnbrough, J., Liu, D., & Flewitt, P. (2013). Micro-scale testing of ductile and brittle cantilever beam specimens in situ with a dual beam workstation. *Measurement Science and Technology, 24*(5), 055010.
24. Mayer, J., et al. (2007). TEM sample preparation and FIB-induced damage. *MRS Bulletin, 32*(5), 400–407.
25. Williams, D. B., & Carter, C. B. (2009). *Transmission electron microscopy: A textbook for materials science*. Boston: Springer US.
26. Liu, M., & Cowley, J. M. (1994). Structures of carbon nanotubes studied by HRTEM and nanodiffraction. *Ultramicroscopy, 53*(4), 333–342.
27. Hofer, F., et al. (2016). Fundamentals of electron energy-loss spectroscopy. In *IOP conference series: Materials science and engineering*. Bristol: IOP Publishing.
28. Demoncy, N., et al. (1998). Filling carbon nanotubes with metals by the arc-discharge method: The key role of sulfur. *The European Physical Journal B-Condensed Matter and Complex Systems, 4*(2), 147–157.
29. Van der Heide, P. (2011). *X-ray photoelectron spectroscopy: An introduction to principles and practices*. Hoboken: Wiley.
30. Haque, M. A., & Saif, T. (2008). Mechanical testing at the micro/nanoscale. In W. N. Sharpe (Ed.), *Springer handbook of experimental solid mechanics*. Boston: Springer US.
31. Cree, D., & Pugh, M. (2011). Dry wear and friction properties of an A356/SiC foam interpenetrating phase composite. *Wear, 272*(1), 88–96.

32. Suresh, K., et al. (2013). Sliding wear behavior of gas tunnel type plasma sprayed Ni-based metallic glass composite coatings. *Vacuum, 88*, 114–117.
33. Abbas, A., et al. (2020). Tribological effects of carbon nanotubes on magnesium alloy AZ31 and analyzing aging effects on CNTs/AZ31 composites fabricated by stir casting process. *Tribology International, 142*, 105982.
34. Huang, S. J., Abbas, A., & Ballóková, B. (2019). Effect of CNT on microstructure, dry sliding wear and compressive mechanical properties of AZ61 magnesium alloy. *Journal of Materials Research and Technology, 8*(5), 4273–4286.
35. Lim, S., et al. (1999). The tribological properties of Al–Cu/SiCp metal–matrix composites fabricated using the rheocasting technique. *Journal of Materials Processing Technology, 89*, 591–596.
36. Rodrigo, P., et al. (2009). Microstructure and wear resistance of Al–SiC composites coatings on ZE41 magnesium alloy. *Applied Surface Science, 255*(22), 9174–9181.
37. Lim, C., Lim, S., & Gupta, M. (2003). Wear behaviour of SiCp-reinforced magnesium matrix composites. *Wear, 255*(1–6), 629–637.
38. Narayanasamy, P., & Selvakumar, N. (2017). Tensile, compressive and wear behaviour of self-lubricating sintered magnesium based composites. *Transactions of Nonferrous Metals Society of China, 27*(2), 312–323.
39. Dash, D., et al. (2020). Influence of TiC on microstructure, mechanical and wear properties of magnesium alloy (AZ91D) matrix composites. *Journal of Scientific & Industrial Research, 29*, 164–169.
40. Thirugnanasambandham, T., et al. (2019). Experimental study of wear characteristics of Al2O3 reinforced magnesium based metal matrix composites. *Materials Today: Proceedings, 14*, 211–218.
41. ASTM. (2017). *G99-17 standard test method for wear testing with a pin-on-disk apparatus*. West Conshohocken: ASTM International.
42. Oliver, W. C., & Pharr, G. M. (1992). An improved technique for determining hardness and elastic modulus using load and displacement sensing indentation experiments. *Journal of Materials Research, 7*(6), 1564–1586.
43. Laha, T., & Agarwal, A. (2008). Effect of sintering on thermally sprayed carbon nanotube reinforced aluminum nanocomposite. *Materials Science and Engineering A, 480*(1–2), 323–332.
44. Duarte, M., et al. (2020). Nanomechanical characterization of a metal matrix composite reinforced with carbon nanotubes. *AIMS Materials Science, 7*(1), 33–45.
45. Chen, J., & Bull, S. (2006). On the relationship between plastic zone radius and maximum depth during nanoindentation. *Surface and Coatings Technology, 201*(7), 4289–4293.
46. Chen, J. (2012). Indentation-based methods to assess fracture toughness for thin coatings. *Journal of Physics D: Applied Physics, 45*(20), 203001.
47. ASTM *E384-17 standard test method for microindentation hardness of materials*. West Conshohocken: ASTM International, 2017.
48. ASTM *E399-19 standard test method for linear-elastic plane-strain fracture toughness KIc of metallic materials*. West Conshohocken: ASTM International, 2019.

Chapter 9
Interface Characterization

Abstract The study of the interfacial phenomena between the reinforcement and the matrix of metal matrix composites is of great importance. These interfacial phenomena have a direct influence on the mechanical properties of composites, since the load transfer between the metallic matrix and the reinforcement depends on the behavior of the interface. It is necessary to understand different aspects such as the chemical stability of the reinforcement in the matrix, the thermodynamic and kinetic aspects involved at the interface, wettability, and the influence of the composite manufacturing process on the matrix-reinforcement interface. This chapter presents and discusses results on such interface in composites of aluminum and magnesium reinforced with MWCNTs produced by the sandwich technique. TEM, HRTEM, and EELS techniques allowed for an in-depth analysis of the matrix-reinforcement interface.

9.1 Introduction to the Interface Behavior Between Metal Matrix and Reinforcement

Interfacial phenomena and chemical stability of the reinforcement in a metal matrix are issues of great importance to understand the mechanical behavior in a metallic composite. For having a good mechanical behavior, the stress must be transferred from the metallic matrix to the reinforcement, i.e., the interface layer between the matrix and the reinforcement must act as bridging for the load transfer [1]. Under these considerations, a strong interface would make the composite very strength, but at the expense of its ductility [2]. A weak interface would lead to lower strength of the composite, facilitating pull-out phenomena at low loads due to the interface failure.

Therefore, the stress transfer between matrix and reinforcement is due to the wettability, which is the interaction of the surface energies; this parameter is represented by the Young's equation (Eqs. 9.1 and 9.2).

$$\cos\theta = \frac{\gamma_{sv} - \gamma_{ls}}{\gamma_{lv}} \tag{9.1}$$

$$W_A = \gamma_{lv}\left(1+cos\theta\right) \tag{9.2}$$

where θ is the contact angle and γ_{sv}, γ_{ls}, and γ_{lv} are the surface energies of solid-vapor, liquid-solid, and liquid-vapor, respectively. Finally, W_A is the adhesion work between the liquid and the substrate.

For a good wettability between the liquid matrix and CNTs, a low liquid-vapor angle is required. Interfacial reactions can allow a phase formation, which, in some cases, can improve wetting if the liquid has a lower contact angle with the phase formed owing to the reaction, and, however, in other cases, the phase formation (interphases) at the interface can allow poor mechanical properties. Several studies has shown the interfacial reactions and degree of wetting of the fibers in a metallic matrix and the response on the composite properties [3–5]. Although the previous model is originally applied to the processing of composites by liquid metallurgy, the principles are also applied to other processing methods.

For aluminum reinforced with carbonaceous materials, the formation of aluminum carbide (Al_4C_3) can be observed at the matrix-reinforcement interface. Some authors have tried to improve the wettability between aluminum and carbon fibers by silicon infiltration; silicon alloys increase the wettability angle, i.e., reduce the surface tension [4]. Vidal-Setif et al. have shown a reduction in strength and premature failure in A357 alloy reinforced with carbon fibers due to the formation of Al_4C_3, as well as the presence of brittle Si particles at the interface [6]. However, there have been reports on property improvement of Al-SiCp composites due to limited amounts of Al_4C_3, suggesting a good interfacial interaction with aluminum [7–9].

For magnesium alloys reinforced with carbonaceous materials, a few studies on the interfacial bonding have been conducted. Yuan et al. [10] studied magnesium-based composites, developing a novel method to increase the interfacial bonding strength by coating magnesium oxide (MgO) nanoparticles on the surface of CNTs. For this kind of composites, Ryou et al. [11] studied the absorption behavior of carbon atoms on the surface of MgO, finding that carbon atoms or graphene could be stably absorbed on the surface of MgO. CNTs possess properties similar to those of graphene sheets. Therefore, a molecular attraction and repulsion model could be established to illustrate the strengthening mechanism of the interfacial bonding between CNTs and MgO [10].

For manufacturing magnesium reinforced with some carbonaceous material, a high temperature is required, causing an interfacial diffusion bonding between the reinforcement and the matrix [12]. Yuan et al. evidenced that the diffused layer formed between CNTs and MgO has a thickness of <2.1 nm, stating that the phase formation at the interface has a strong bonding [10]; however, the behavior at the magnesium-CNTs interface shows an incoherent interface.

Notwithstanding, other studies have shown good mechanical behavior at the interface in magnesium composites. Li et al. [13] reported good mechanical properties of Mg-CNT composites due to the good interfacial bonding at the interface;

additionally, the absence of nanopores at the interface allowed a good interaction matrix-reinforcement, which was also reported by Shi et al. [14].

Carbide formation between carbonaceous reinforcements and metallic matrices occurs by nucleation and growth phenomena. Concerning the nucleation processes, there is a critical dimension associated with the carbide formation [15], which is given by Eq. 9.3.

$$t_{crit} = V_m \frac{\Delta \gamma}{\Delta G^f} \quad (9.3)$$

where V_m is the molar volume of the carbide formed, ΔG^f is the free energy of formation per mole of carbide, and $\Delta \gamma = \gamma_{MC/CNT} + \gamma_{MC/Alloy} - \gamma_{alloy/CNT}$ is the increase in the total surface as a result of the formation of a new interface; MC refers to metal carbide. If the thickness of the carbide reaches the critical thickness, it would dissolve in the molten metal. Besides, the carbide formation could decrease the contact angle and improve the wettability between metal matrix and reinforcement.

9.2 Interface Strengthening Mechanisms in Light Metallic Materials

The prediction capability of the existing models [16–20] depends on how the matrix interacts with the reinforcing material. These interactions are influenced, among other factors, by the wettability of the reinforcement, interface strength, and residual stresses.

The stress transmission in fiber-reinforced composites can occur through the following external loads. If the reinforcement is in the direction perpendicular to the load, the stress transmission will be through the interface, whereby the mechanical properties will depend on the permissible stresses (i.e., the shear stress) of the interfacial material. It is worth mentioning that, if there is no phase formation, the anchoring could be through weak attractive forces between reinforcement and matrix; this kind of load transfer mainly occurs when the metallic matrix is reinforced with single-walled carbon nanotubes (SWCNTs) [21]. For a specific kind of reinforcement, such as CNTs, if it is aligned in the loading direction, the stress transfer can occur through two mechanisms. The first is at the CNT ends and the second at the CNT surfaces. In these mechanisms, the composite stresses are transmitted through interface shear stress. The property prediction by this mechanism is shown in Eq. 9.4 [22], which takes into account the phase formation at the metal matrix-CNTs interface.

$$\sigma_c = \left(\frac{1+2b}{D}\right)\left[\frac{\sigma_{shear} l}{D} - \left(\frac{1-2b}{D}\right)\sigma_m\right]v_f + \sigma_m \quad (9.4)$$

where σ_c is the composite stress, σ_{shear} is the shear stress at the interface (which is not easy to measure), σ_m is the maximum stress in the matrix, b is the interface thickness, D is the CNT diameter, and v_f is the volume fraction of the fiber in the composite.

The interactions between CNTs and the metal matrix can be given in two zones of the CNTs: at the ends (two faces) and at the longitudinal face of their surface (one face). At the end zone, sigma-like links to the matrix by coherent interface formation can be developed. At the longitudinal face of the surface, the bonds can also be coherent or weaker. The presence of some defects to promote the interphase formation is needed, which finally will allow a good transfer of stresses.

Another important behavior that may occur during the manufacturing process is the dislocation stacking on the reinforcement surface. The formation of dislocation forest at the interface can cause strengthening and is represented by the Taylor expression shown in Eq. 9.5 [23].

$$\sigma = \sigma_0 + \alpha M^T G b \rho^{1/2} \qquad (9.5)$$

where σ is the flow stress, σ_0 is the friction stress, α is a constant (1/3), M^T is the Taylor factor (3 for untextured polycrystalline materials), G is the shear modulus, b is the Burgers vector of the dislocations, and ρ is the dislocation density. Additionally, the dislocations in the regions near the interface can be affected by thermal effects [24] and by dislocation loops that may be generated during the manufacturing process [25].

9.3 Other Interactions Between CNTs and Metal Matrix

Surface roughness of the reinforcement has a very important role on the matrix-reinforcement interface. The surface roughness can allow the mechanical transfer of the loads through a "hook" effect, but can modify the wettability of the surface.

Another important aspect to consider is the chemical or physical functionalization. An appropriate functionalization allows good CNT anchoring to the matrix. In this way, a good interface allows an adequate stress transfer to improve the mechanical properties. The chemical functionalization is based on the covalent bonding among functional carbons at the CNT surface or CNT ends. The functionalization is associated with a change in hybridization of the sp^2 to sp^3 orbitals and a simultaneous loss of π-conjugations of the graphene layers [26]. A study on aluminum composites [27] shows amino-functionalization as a way to generate strong bonds between CNTs and the aluminum matrix. Currently it is possible to buy functionalized CNTs in considerable amounts and inexpensive prices.

9.4 Interface Characterization in Metal Matrix Composites

As an exercise of the characterization in metal matrix composites, the interface behavior in aluminum and magnesium matrices reinforced with MWCNTs will be discussed. While these composites were fabricated by the sandwich technique [28–30] described in Chap. 4, the analysis can be applied to different composites produced by other routes.

For the interface and interphase characterization between the metallic matrix and MWCNTs, several microscopy techniques were used. Dislocation stacking and coherent or semi-coherent structure between the metallic matrix and MWCNTs were found. However, a detailed analysis at the interface was done for magnesium composites.

9.4.1 Interface Characterization in Composites Fabricated by Sandwich Technique

In order to conduct a comprehensive study between metallic sheets, transversal sections of the reinforced zones of some samples were prepared by a focused ion beam (FIB) system, as shown in Fig. 9.1.

An elemental analysis in different zones close to the interface between metallic sheets (diffusion zone) was done by energy-dispersive spectroscopy (EDS) coupled to the scanning electron microscope (SEM). The results for both composites are shown in Fig. 9.2.

For aluminum composites (Fig. 9.2a), two zones were identified: zone with high MWCNT content and zone with pure aluminum. The former shows carbon and aluminum as main elements, in addition to nickel and oxygen. Nickel comes from the catalyst used in the MWCNT synthesis, while oxygen can come either from the

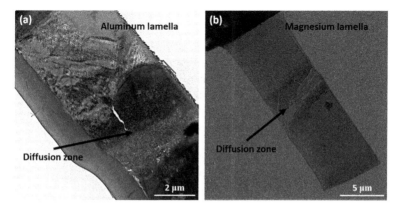

Fig. 9.1 FIB preparation of lamellae: (**a**) aluminum and (**b**) magnesium composites

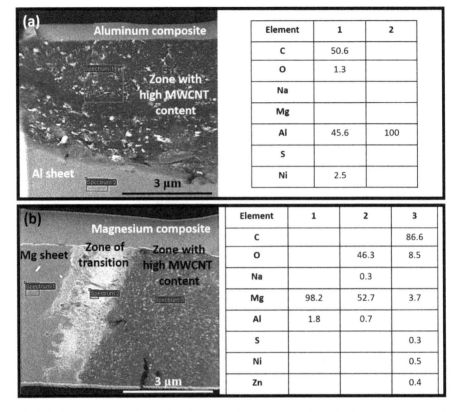

Fig. 9.2 SEM-EDS analysis (wt%) performed at different zones of (**a**) aluminum composites and (**b**) magnesium composites

polyvinyl alcohol degradation or from the oxidation of aluminum during the manufacturing process (Chap. 4).

Regarding magnesium composites, Fig. 9.2b presents the results of three spectra taken. Spectrum 1 (Mg sheet) shows magnesium and aluminum as main elements; this zone was not affected by the diffusion process during the composite manufacturing. Spectrum 2 (transition zone) displays a high amount of oxygen; as will be verified below by a TEM diffraction pattern, the oxygen reacted with magnesium forming magnesium oxide (MgO). Spectrum 3 (high MWCNT content) was taken at the interface zone between magnesium layers, where the main elements are carbon, magnesium, and oxygen; nickel comes from the MWCNT synthesis.

9.4 Interface Characterization in Metal Matrix Composites 173

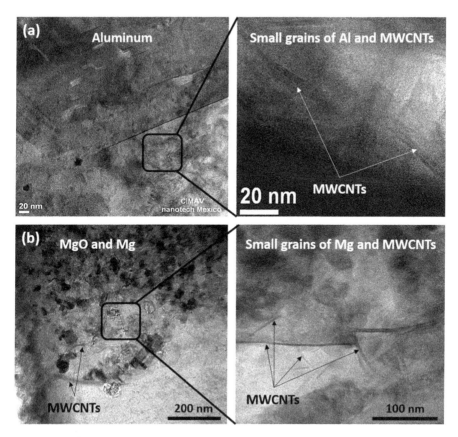

Fig. 9.3 TEM bright field images for composites: (**a**) aluminum and (**b**) magnesium reinforced with 0.5 wt% MWCNTs

9.4.2 TEM Analysis of Metal Matrix Composites

For both aluminum and magnesium composites, TEM images were acquired at the interface between metallic sheets, i.e., at the diffusion zone shown in Fig. 9.1. For the aluminum composite (Fig. 9.3a), it is possible to evidence MWCNTs well embedded into the metallic matrix; no phase formation was identified at the interface. For the magnesium composite (Fig. 9.3b), MWCNTs, magnesium (bright phase), and magnesium oxide (dark phase) were identified; the close-up image shows MWCNTs well embedded in both Mg and Mg/MgO zones. Further, it is possible to see small grains around the MWCNTs, i.e., there was a recrystallization caused by the high plastic deformation and temperature during the manufacturing process.

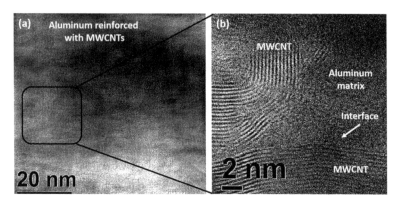

Fig. 9.4 (a) TEM and (b) HRTEM images of aluminum reinforced with 0.5 wt% MWCNTs

9.4.3 HRTEM Analysis at the Interface Between MWCNTs and Metal Matrix

TEM observation confirmed that MWCNTs were singly dispersed and completely embedded in both aluminum and magnesium composites. No intermediate compounds or nanopores were detected at the interface between MWCNTs and the metal matrix, which indicates that a good interfacial bonding was achieved. Then, HRTEM images were obtained close to the interface between MWCNTs and the metal matrix for both aluminum and magnesium composites. As will be discussed below, some strengthening mechanisms were identified, which allowed to explain the improvement in mechanical properties with the addition of MWCNTs.

TEM and HRTEM images for the aluminum reinforced with 0.5 and 2 wt% MWCNTs are shown in Figs. 9.4 and 9.5, respectively. The images confirmed that the MWCNTs are well embedded into the aluminum matrix, with no evidence of cluster formations. As already known [31, 32], these features are essential for a good load transfer and thereby to improve the mechanical properties of metal matrix composites. These images did not show evidence of MWCNTs damage. Images in Fig. 9.5 allowed to measure the interplanar distance (0.23 nm) of the matrix (Fig. 9.5b), which corresponds to the plane (1 1 1) of the aluminum (ICDD database PDF-2/2003), as well as the interlayer spacing (0.33 nm) of the MWCNTs (Fig. 9.5c). In the evaluated zones, HRTEM did not show the presence of the harmful aluminum carbide phase, which confirms the findings of X-ray diffraction (XRD) reported in Chap. 4. With this, a good load transfer from the matrix to the reinforcement is achieved.

For magnesium reinforced with different MWCNT content, a detailed analysis was done; TEM and HRTEM images are shown in Figs. 9.6, 9.7, and 9.8. The images evidence that a well-graphitized MWCNT structure was retained in the composites, which indicates that the MWCNTs were not damaged during the whole composite manufacturing process. Dislocation stacking (Fig. 9.6) and semi-coherent

9.4 Interface Characterization in Metal Matrix Composites

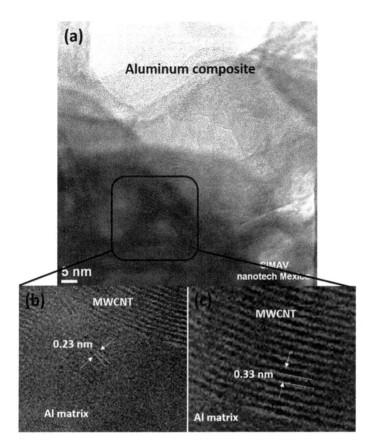

Fig. 9.5 (**a**) TEM and (**b**) and (**c**) HRTEM images of aluminum reinforced with 2 wt% MWCNTs

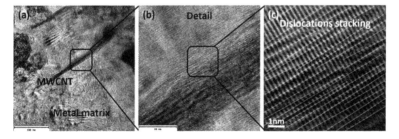

Fig. 9.6 HRTEM images for magnesium reinforced with 0.25 wt% MWCNTs. The dislocation stacking can be seen in (c) (HRTEM filter)

and coherent interfaces (Figs. 9.7 and 9.8) were identified in some zones at the interface between MWCNTs and magnesium matrix.

Bright and dark field images by the annular dark-field (ADF) technique are shown in Fig. 9.7. The close-up images (details 1 and 2) reveal that the metal matrix and the MWCNTs have a coherent interface, at least in that study zone. Fast Fourier

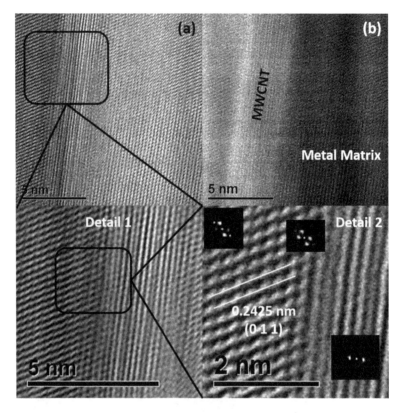

Fig. 9.7 HRTEM images of magnesium reinforced with 0.5 wt% MWCNTs taken at the interface zone (HRTEM filter)

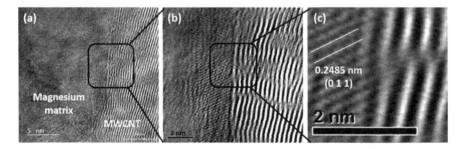

Fig. 9.8 HRTEM images of magnesium reinforced with 1.0 wt% MWCNTs. Coherence of the interface between MWCNTs and metal matrix is shown (HRTEM filter)

transform (FFT) images (detail 2), which represent the diffraction pattern corresponding to a mixture of magnesium and MWCNTs, show how the crystalline structure changes between the metal matrix and the MWCNT. Only the presence of magnesium and MWCNT are detected, with no other compound or interphase close to the interface between magnesium and MWCNTs. The region close to the

9.4 Interface Characterization in Metal Matrix Composites

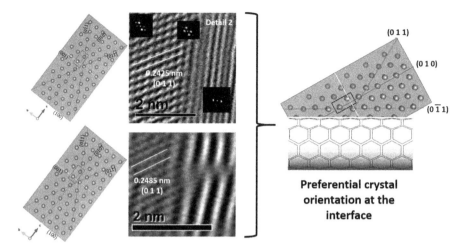

Fig. 9.9 Orientation analysis at the interface between MWCNTs and magnesium matrix

MWCNTs is comprised mainly of (0 1 1) planes and has an interplanar distance of 0.2425 nm, which corresponds to the magnesium phase. The interface image suggests the formation of a coherent interface between MWCNT and metal matrix, with which the load can be transferred from the metal matrix to MWCNTs, in such a way that the reinforcement effectively acts in the composite. After indexing the crystallographic planes in the high-resolution images, it is evident that the interface between MWCNTs and metal matrix has a preferential orientation (texturing); this analysis will be detailed later.

Figure 9.8 shows HRTEM images for the composite reinforced with 1.0 wt% MWCNTs. A similar behavior to that of the composites with 0.25 and 0.5 wt% MWCNTs was found. Moreover, the study zone in the composite reinforced with 1.0 wt% MWCNTs evidences that the {0 1 1} planes family and (0 1 0) plane appear at the surface of the Mg matrix and have a similar orientation to that of the composites with 0.25 and 0.5 wt% MWCNTs.

The relative orientation of the planes near the interface with respect to the MWCNT surface was investigated using VESTA software [33]. For this analysis, an hcp unit cell was simulated, and, with the interplanar distance measured in HRTEM images, the orientation of the unit cell was identified. Figure 9.9 reveals that the {0 0 0 2} basal planes, with an angle of 45° with respect to the MWCNT walls, are the preferred ones close to the interface between MWCNTs and Mg matrix.

The findings presented above are paramount because the plastic deformation of magnesium at room temperature is limited to two main deformation mechanisms: $\{0001\}$ <a> - $11\bar{2}0$ basal planes and mechanical twinning. Other systems such as ($10\bar{1}0$) prismatic slip and pyramidal slip have in common the same <a> - $11\bar{2}0$ slip direction, but require larger critical resolved shear stresses for activation. Accordingly with the literature [34], under loading in directions either parallel or perpendicular to the sheet plane, basal, prismatic and pyramidal, the <a> slip systems fail to accommodate any deformation because they all have a slip direction parallel to the basal plane, and the resulting shear strain in all slip systems is zero. For such cases, slip vectors with a component out of the basal plane are required. For the investigated systems (MWCNT and magnesium matrix), the basal plane is at an angle of 45° with respect to the MWCNTs walls, as shown in Fig. 9.9. Because the plastic deformation at the interface of the MWCNT walls is difficult to perform, the load transfer between the metal matrix and the MWCNTs will be good. In order to activate other slip systems, a new vector acting on the pyramidal plane in the $\{11\bar{2}3\}$ direction is necessary, which is referred to as pyramidal <c + a> slip. This slip mode offers five independent slip systems, thus satisfying the von Mises criterion for an arbitrary shape change. Therefore, pyramidal <c + a> slip in magnesium usually requires substantial thermal activation, i.e., high temperatures or high shear stress for activation. With the absence of <c + a> slip at low deformation temperatures, the hexagonal crystals have no ways to accommodate the imposed strain along the sheet normal direction by crystallographic slip. This causes high stresses, modest work hardening, and, most importantly, premature brittle fracture. Prismatic slip is not common in magnesium, but <c + a> dislocation slip on pyramidal planes is possible. The relative difficulty of these "harder" slip modes compared to the "softer" ones plays an important role in determining textures, strength, and ductility.

The ease of slipping on different crystal systems is usually quantified by their critical resolved shear stresses (CRSSs), which are measured in Stage I deformation on unconstrained single-crystal specimens where a unique slip system is activated. The values of stress for these planes and generated strain in the structure are 0.81 MPa for the basal plane, 45 MPa for prismatic slip planes, and 80 MPa for pyramidal plane in a single crystal. These crystal orientations allow to identify the shear stress at the interface between metal matrix and MWCNTs, as well as to predict the theoretical stress in the composite according to the kind of interface, i.e., according to the crystal hcp orientation. On the other hand, this preferential orientation allows a very good load transfer between magnesium and MWCNTs due to the few plastic deformation that this direction in hexagonal structure has; during the stress flow through the interface, few dislocations could appear and the stress transfer is effective.

In summary, HRTEM analysis allowed to identify the interactions between the metal matrix and MWCNTs. Both semi-coherent and coherent interaction between the metal matrix and MWCNTs were found by HRTEM images. Additionally, the stacking and formation of dislocations were identified at the interface, and no intermediate interphases were found in the studied zones for both aluminum and magnesium composites. Finally, the interface between metal matrix and MWCNTs has a preferential growth direction, as shown in Fig. 9.9, and the basal planes at 45° with respect to the MWCNT walls are preferential close to their interface. This orientation allows to know the shear stress at the interface between MWCNTs and metal matrix. Finally, with the elucidation of all mechanisms occurring at the interface, a reliable prediction of the composite behavior is possible to do.

9.4.4 Elemental Analysis and Energy Loss Spectroscopy at the Interface Between MWCNTS and Metal Matrix

Energy-dispersive spectroscopy (EDS) and electron energy loss spectroscopy (EELS) are used to identify the chemical elements in alloys and composites, as well as the chemical interaction between the metal matrix and reinforcements.

Figure 9.10 shows a TEM-EDS elemental analysis for magnesium reinforced with 0.5 wt% MWCNTs, achieved at the magnesium matrix-MWCNT interface; Mg, Al, C, O, and Zn were identified. The EDS elemental mapping presented in Fig. 9.11 allows to see how the elements are distributed along the analyzed area. As stated before for the SEM-EDS results, the oxygen comes from the PVA thermal

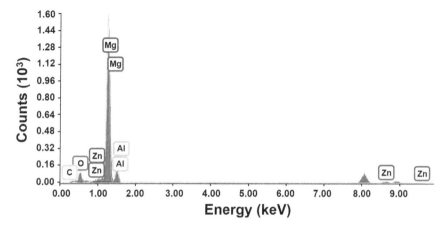

Fig. 9.10 TEM-EDS elemental analysis of magnesium reinforced with 0.5 wt% MWCNTs

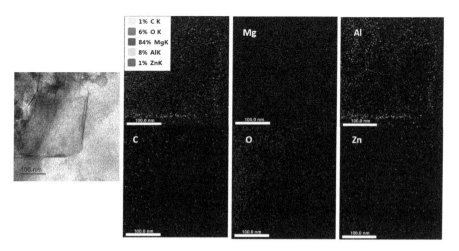

Fig. 9.11 TEM-EDS elemental mapping of magnesium reinforced with 0.5 wt% MWCNTs

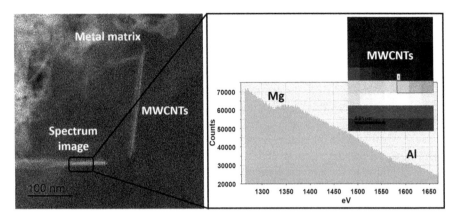

Fig. 9.12 EELS analysis at the metal matrix-MWCNTs interface of a Mg-MWCNTs composite

degradation during the hot compaction process of the sandwich technique (Chap. 4), while the carbon comes from the MWCNTs.

Besides, the aluminum diffuses to the MWCNT walls, which is due to the chemical affinity between aluminum and carbon for forming aluminum carbide by hot processes. However, the Al-C diffusion pair is not enough to form some interphase at the metal matrix-MWCNT interface for both aluminum and magnesium composites studied. Indeed, the results of EELS (Fig. 9.12) suggest the presence of only aluminum atoms close to the MWCNT walls. Further, the variation of free energy of carbide formation shows a positive value for magnesium carbide; thus, the formation of this kind of compounds is unlikely to happen in these composites.

9.5 Conclusions

The sandwich technique has proven to be an effective technique for manufacturing light metal matrix composites. The interface characterization of both aluminum and magnesium composites was carried out by TEM and HRTEM techniques. The results showed that the MWCNTs are well dispersed and embedded into the metal matrix. MWCNTs did not present damage or morphological changes, such as their diameter and length. A strong and good interface between the MWCNTs and the metal matrix was evidenced. Interphases were not found in the studied zones. In the case of magnesium composites, while some reactions between MWCNTs and aluminum particles were evidenced, there was no carbide formation. The presence of MWCNTs, acting as second phases into the metal matrices, led to an increase in the work hardening. MWCNTs drive to increased nucleation rates during recrystallization processes, leading a fine grain structure as was shown by TEM images. On the other hand, dislocation strengthening is possible due to the dislocation pile-up against the MWCNT external walls.

References

1. Coleman, J. N., et al. (2006). Small but strong: A review of the mechanical properties of carbon nanotube–polymer composites. *Carbon, 44*(9), 1624–1652.
2. Dilandro, L., Dibenedetto, A., & Groeger, J. (1988). The effect of fiber-matrix stress transfer on the strength of fiber-reinforced composite materials. *Polymer Composites, 9*(3), 209–221.
3. Piggott, M. (1989). The interface in carbon fibre composites. *Carbon, 27*(5), 657–662.
4. Chen, H., & Alpas, A. (1996). Wear of aluminium matrix composites reinforced with nickel-coated carbon fibres. *Wear, 192*(1-2), 186–198.
5. Poteet, C., & Halla, I. (1997). High strain rate properties of a unidirectionally reinforced C/Al metal matrix composite. *Materials Science and Engineering A, 222*(1), 35–44.
6. Vidal-Setif, M., et al. (1999). On the role of brittle interfacial phases on the mechanical properties of carbon fibre reinforced Al-based matrix composites. *Materials Science and Engineering A, 272*(2), 321–333.
7. Laha, T., et al. (2007). Interfacial phenomena in thermally sprayed multiwalled carbon nanotube reinforced aluminum nanocomposite. *Acta Materialia, 55*(3), 1059–1066.
8. Tham, L., Gupta, M., & Cheng, L. (2001). Effect of limited matrix–reinforcement interfacial reaction on enhancing the mechanical properties of aluminium–silicon carbide composites. *Acta Materialia, 49*(16), 3243–3253.
9. Kwon, H., et al. (2009). Combination of hot extrusion and spark plasma sintering for producing carbon nanotube reinforced aluminum matrix composites. *Carbon, 47*(3), 570–577.
10. Yuan, Q.-h., et al., Microstructure and mechanical properties of AZ91 alloy reinforced by carbon nanotubes coated with MgO. Carbon, 2016. *96*: p. 843-855.
11. Ryou, J., & Hong, S. (2013). First-principles study of carbon atoms adsorbed on MgO (100) related to graphene growth. *Current Applied Physics, 13*(2), 327–330.
12. Wan, X.-J., & Lin, J.-G. (2011). Interfacial microstructure and chemical stability during diffusion bonding of single crystal Al2O3-fibres with Ni25. 8Al9. 6Ta8. 3Cr matrix. *Transactions of Nonferrous Metals Society of China, 21*(5), 1023–1028.
13. Li, C., et al. (2014). Microstructure and strengthening mechanism of carbon nanotubes reinforced magnesium matrix composite. *Materials Science and Engineering A, 597*, 264–269.
14. Shi, H., et al. (2014). A novel method to fabricate cNT/Mg–6Zn composites with high strengthening efficiency. *Acta Metallurgica Sinica (English Letters), 27*(5), 909–917.
15. Landry, K., et al. (1997). Mechanisms of reactive wetting: the question of triple line configuration. *Acta Materialia, 45*(7), 3079–3085.
16. Zhang, Z., & Chen, D. (2008). Contribution of Orowan strengthening effect in particulate-reinforced metal matrix nanocomposites. *Materials Science and Engineering A, 483*, 148–152.
17. Zhang, Z., & Chen, D. (2006). Consideration of Orowan strengthening effect in particulate-reinforced metal matrix nanocomposites: A model for predicting their yield strength. *Scripta Materialia, 54*(7), 1321–1326.
18. Sanaty-Zadeh, A. (2012). Comparison between current models for the strength of particulate-reinforced metal matrix nanocomposites with emphasis on consideration of Hall–Petch effect. *Materials Science and Engineering A, 531*, 112–118.
19. Nardone, V., & Prewo, K. (1986). On the strength of discontinuous silicon carbide reinforced aluminum composites. *Scripta Metallurgica, 20*(1), 43–48.
20. Hull, D., & Bacon, D. J. (2011). *Introduction to dislocations* (5th ed.). New York: Butterworth-Heinemann, Elsevier.
21. Agarwal, A., Bakshi, S. R., & Lahiri, D. (2018). *Carbon nanotubes: Reinforced metal matrix composites*. CRC Press.
22. Coleman, J. N., et al. (2004). High performance nanotube-reinforced plastics: Understanding the mechanism of strength increase. *Advanced Functional Materials, 14*(8), 791–798.
23. McQueen, H., & Kassner, M. (2004). Comments on 'a model of continuous dynamic recrystallization' proposed for aluminum. *Scripta Materialia, 51*(5), 461–465.

24. George, R., et al. (2005). Strengthening in carbon nanotube/aluminium (CNT/Al) composites. *Scripta Materialia, 53*(10), 1159–1163.
25. Orowan, E. (1934). über den mechanismus des gleitvorganges (Mechanisms of sliding process). *Zeitschrift für Physik, 89*, 634.
26. Ma, P.-C., et al. (2010). Dispersion and functionalization of carbon nanotubes for polymer-based nanocomposites: A review. *Composites Part A: Applied Science and Manufacturing, 41*(10), 1345–1367.
27. Singhal, S., et al. (2012). Carbon nanotubes: amino functionalization and its application in the fabrication of Al-matrix composites. *Powder Technology, 215*, 254–263.
28. Isaza, C., Sierra, G., & Meza, J. (2016). A novel technique for production of metal matrix composites reinforced with carbon nanotubes. *Journal of Manufacturing Science and Engineering, 138*(2), 024501.
29. Isaza, M. C. A., et al. (2017). Mechanical properties and interfacial phenomena in aluminum reinforced with carbon nanotubes manufactured by the sandwich technique. *Journal of Composite Materials, 51*(11), 1619–1629.
30. Merino, C. A. I., et al. (2017). Metal matrix composites reinforced with carbon nanotubes by an alternative technique. *Journal of Alloys and Compounds, 707*, 257–263.
31. Jiang, L., et al. (2011). An approach to the uniform dispersion of a high volume fraction of carbon nanotubes in aluminum powder. *Carbon, 49*(6), 1965–1971.
32. Morsi, K., & Esawi, A. (2007). Effect of mechanical alloying time and carbon nanotube (CNT) content on the evolution of aluminum (Al)–CNT composite powders. *Journal of Materials Science, 42*(13), 4954–4959.
33. Momma, K., & Izumi, F. (2011). VESTA 3 for three-dimensional visualization of crystal, volumetric and morphology data. *Journal of Applied Crystallography, 44*(6), 1272–1276.
34. Chen, P. (2019). *Twin-slip interaction in plastic deformation of magnesium*. Reno: University of Nevada.

Chapter 10
Applications in the Aeronautical and Aerospace Industries

Abstract This chapter presents a review of some industrial applications of light alloys and composites. While aluminum alloys and their composites play an important role in the development of aeronautical, aerospace, defense, and automotive industries owing to their improved mechanical properties and low weights, other lightweight materials such as Ti and Mg are being also successfully applied in these industrial sectors. The high-performance characteristics of a modern aeronautic industry are a direct consequence of the high-performance light materials, composites, and their manufacturing. The selection of the alloy type, with characteristics that will depend on the expected application, is a very important aspect from the point of view of its industrial implementation. A case study of the manufacture of thermal spray coatings, based on the use of light alloys with high potential for industrial-scale production, is here presented.

10.1 Introduction to Industrial Applications of Light Alloys and Composites

The driving force for the development of Al alloys for aerospace structures has been the improvement in static strength, as well as fracture toughness and resistance to crack growth, particularly in fatigue regimes, damage tolerance, and stress corrosion. In the same context, the weight saving introduced by magnesium alloys encouraged their widespread and successful use for helicopter components. However in this application, magnesium (Mg) alloys are also very susceptible to surface damage due to impact, which occurs frequently during manufacture and/or overhaul and repair.

Among the aluminum alloys available for mass production, the 6061-T6 alloy is the most widely applied. This alloy has excellent joining characteristics and good acceptance for applied coatings and combines relatively high strength, good workability, and high resistance to corrosion. The most important aeronautical application of this alloy is in aircraft fittings. Besides, the 6XXX series Al alloys containing Mg and Si are used for architectural extrusions and automotive components. The 2XXX series, in which copper is the main alloying element, and 7XXX series, in which zinc is the main alloying element, are at the forefront of defense and aerospace industries. Three specific aluminum alloys, Al-2024, Al-6082, and Al-7075, are the most applied in many aeronautical and aerospace components [1]. Al-2024 alloy has good machinability, high strength, and fatigue resistance. Because of that, it is widely used in aircraft structures, gears and shafts, and missile parts. Al-6082 alloy is known as a structural alloy, which has medium strength and good corrosion resistance. This alloy is used for cranes, highly stressed applications, and transport applications. General characteristics of the Al-7075 alloy are very high strength comparable to many steels, good fatigue strength, and average machinability, but with less resistance to corrosion than many other Al alloys. This alloy is used for highly stressed structural part manufacturing, mainly aircraft and other aerospace applications [2, 3].

Figure 10.1 schematically presents the main factors influencing the needed light alloy and composite properties, planned to fulfill the specific requirements of the intended industrial application. As it could be seen in the previous chapters of this book, the category of composites is an extremely broad definition of different materials and manufacturing processes. All mechanical powder mixtures and matrix/reinforcement can be defined as composites. Also, there exists a number of composite materials, e.g., in the group of abradable coatings. The structure and consequently the technical properties of thermally sprayed coatings are highly dependent on several factors, such as the material composition and class, the feedstock material characteristics, the spray process and parameters used, the coating formation, and the posttreatment procedures. The presence of such a large number of thermal spraying parameters gives huge possibilities for tailoring coating properties and producing excellent coatings with specific properties. However, this also makes the process to some extent sensitive to variations in spray conditions, which may result in the deviation of the coating quality. Nevertheless, thermal spray technology can be a highly reproducible deposition process if all important parameters are monitored and controlled during spraying.

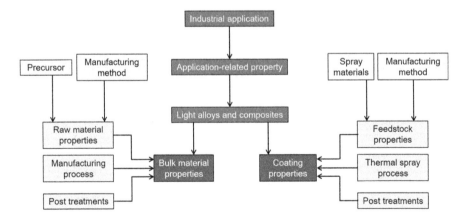

Fig. 10.1 Factors influencing the required properties of light alloys and composites for an industrial application

A case study of the manufacture and application of thermal spray coatings will be presented below. Thermal spray processes are applicable at the industrial level to minimize problems related to microstructure, phase structure, density, mechanical properties, corrosion properties, tribological properties, and high-temperature properties.

10.2 Applications of Thermal Spray Coatings

The growing demand of the industry for having better quality coatings and lower costs, implies the development of new coatings with less surface roughness and better adhesion to the substrates. Thermal spray coatings are used in components and equipment designed to have high performance and durability. Applications in which a thermal spray process is one part of the full manufacturing chain are known. At present, thermal spray technologies are used in many different industrial applications and sectors. For instance, they are used in the refurbishment of worn or incorrectly machined parts; wear protection against abrasion, erosion wear, and adhesive wear; corrosion protection against atmospheric and wet corrosion; protection against oxidation and high-temperature corrosion; and as thermal environmental barriers coatings at high temperatures, e.g., TEBCs. Thermal spray technology is continuously developing, and new applications are seen for thermally sprayed coating materials and structures. Figure 10.2 shows pictures of some of the processes used at the industrial level.

As an example of an industrial application, here we present an experimental case of the use of thermal spray techniques to coat aluminum alloys. The aim is to increase the mechanical and tribological performance of the substrate, which can open new applications in the industrial sectors mentioned above. Two different coatings were applied onto aluminum alloys to improve their surface properties. AlSiC (310NS) and FeCrBSiNbW (140MXC) alloys were deposited on 6061-T6 aluminum alloy, using combustion flame spray and arc wire spray processes, respectively. Several priority features are analyzed, for instance: the metal alloy particle melting; the splat formation, including solid-liquid two-phase droplet impact, which are involved in the coating deposition; cross-sectional microstructure; and bonding with the substrate. The properties of the coatings were investigated according to the coating type, considering mechanical and tribological properties, as well as sliding wear resistance of the sprayed coatings.

The applied coatings were evaluated in the cross section of samples. Vickers microhardness (HV) test was performed using a Clemex MMT-X7 microhardness tester, applying an indentation load of 200 gf for 10 s to produce a series of indentation marks separated by 100 µm. For a more detailed mechanical characterization, the elastic modulus and hardness of the samples were determined by nanoindentation on a TI 950 Triboindenter by Hysitron, equipped with the Berkovich diamond tip. The maximum applied load of 1 mN was used to characterize the softer phases, and a load of 2.5 mN was used for harder phases. All nanoindentations were made

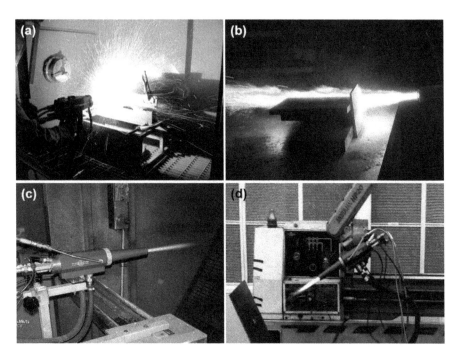

Fig. 10.2 Thermal spray processes used in industrial sectors. (**a**) Arc spray process. (**b**) Combustion flame spray process. (**c**) High-velocity oxy-fuel (HVOF) process. (**d**) Robotized HVOF gun. (Images (**c**) and (**d**) courtesy of Eutectic Mexico)

with a loading-unloading rate of 150 μN/s. Several phases seen on the cross section of samples were analyzed by nanoindentation technique, producing at least ten nanoindentations per phase area according to the established measurement technique. The resulting load-displacement (P-h) curves were used to calculate the elastic modulus and hardness of the phases, according to the Oliver and Pharr method [4]. Friction and wear tests were performed on a pin-on-disk CSM tribometer equipped with software Tribox 4.1. An Al_2O_3 ball of 6 mm in diameter was used as a counter body, and the tests were performed at a normal load of 5 N. The sliding speed was 0.1 m/s with a sliding distance of 100 m. The room temperature was measured as 25 ± 1 °C with relative humidity of 25–30%. Before the wear tests, the coating samples were cleaned with ethanol. The worn surfaces were characterized by SEM. Wear resistance was reported as the volume loss according to the ASTM G99-17 standard [5] considering an average of 10 measurements. The disk volume loss in mm^3 was calculated according to Eq. 10.1; the wear track (ǔ37;) was measured using image analysis with the Image-Pro Plus software. The width of the wear track was automatically evaluated with this software based on grayscale delimitation, considering an average of ten images previously acquired with an optical microscope at 100X. The coating wear rate in mm^3/Nm was calculated using Eq. 10.2.

10.2 Applications of Thermal Spray Coatings

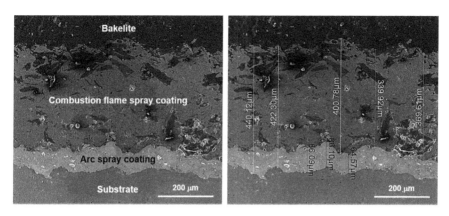

Fig. 10.3 SEM image of bilayer coating cross section deposited first by the AS process and then by the CFS process on the 6061-T6 aluminum alloy

$$V = \frac{(\pi RD^3)}{6r} \quad (10.1)$$

where ǔ49; is the loss volume (mm³), ǔ45; is the wear track radius (mm), ǔ5F; is the sphere radius (mm), and ǔ37; is the wear track (mm).

$$W_{RC} = \frac{V}{(LS)} \quad (10.2)$$

where ǔ4A;$_{RC}$ is the coating wear rate, ǔ3F; is the load (ǔ41;), and ǔ46; the test length (m).

Figure 10.3 shows the morphology and microstructure of the cross-sectional bilayer coating, which was homogeneously deposited firstly by the arc spray (AS) process and then by the combustion flame spray (CFS) process. This sample was manufactured with the aim of observing physical compatibility between two different processes, which could open new applications for 6061-T6 aluminum alloys, taking advantage of each chemical composition and properties. In this case, both coatings are adhered and do not present substrate detachments or additional interfaces.

Figure 10.4 presents a comparison of the microhardness results of thermal spray coatings deposited on the 6061-T6 aluminum alloy. The corresponding average microhardness of the substrate is around 110 ± 12 HV and, due to the T6 heat treatment, is slightly higher than that of the 310NS coating (85 ± 18 HV), which was deposited by combustion flame spray process. However, the microhardness value is very high for the 140MXC coating deposited by the arc spray process. In this case, the coating reaches an average microhardness of 1017 ± 76 HV, which is evident due to the very different chemical composition and phases involved in the process. This hardness value is similar to that of the bulk Fe-based metallic glass material

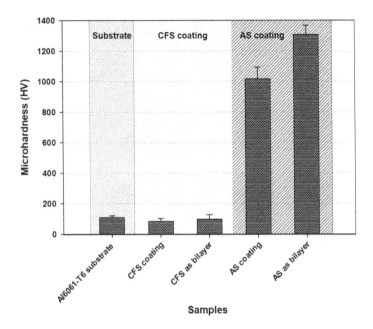

Fig. 10.4 Bulk Vickers microhardness results from specific phases for both types of coatings deposited on the 6061-T6 aluminum alloy. CFS, combustion flame spray process; AS, arc spray process

and as-sprayed metallic glass coating measured in the $Fe_{0.75}B_{0.2}Si_{0.06}Nb_4$ bulk glassy alloy (1070 HV) using the same testing conditions [6]. It is interesting to notice that a higher microhardness value (1307 ± 61 HV) was obtained for the sample with the 140MXC coating deposited as a bilayer by arc spray process, under the same process parameters. This mechanical behavior agrees with the phase distribution and coating morphology observed previously in Fig. 10.3. In the combustion flame spray process, the carbon phase, which is more fragile, migrated to the surface, causing a contraction of the arc spray coating thickness and therefore hardening the bulk material. The reason to obtain a high microhardness in the arc spray coating can be interpreted as follows. The low porosity quantity obtained in the coating, contributes to an increase in its hardness. The amorphous/nanocrystalline alloy coating presents a denser structure with higher elastic properties, which is favorable to the formation of higher residual stresses and an increased hardness. On the other hand, the microstructure of the coating consists of an amorphous phase matrix with nanoparticles well embedded and distributed. The very hard carbide particles act as the main reinforcement phase in the nanocomposite, and therefore the bulk microhardness is higher.

For a more refined study, nanoindentation analyses of both types of coatings were performed in specific main phases. The results of hardness and reduced modulus are presented in Fig. 10.5. For the sample deposited with the 310NS coating by the combustion flame spray process, the AlSi matrix has a better reduced modulus

10.2 Applications of Thermal Spray Coatings

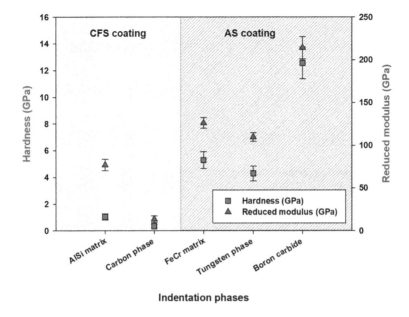

Fig. 10.5 Nanoindentation results from specific phases for both types of coatings deposited on 6061-T6 aluminum alloy. CFS, combustion flame spray process; AS, arc spray process

(E_r = 77 ± 6.6 GPa) compared to the carbon phase (E_r = 14 ± 3.7 GPa). However, the hardness of these phases does not vary greatly, presenting values of 1.0 ± 0.2 GPa and 0.4 ± 0.1 GPa, respectively. On the other hand, for the sample deposited with the 140MXC coating by the arc spray process, the boron carbide phase has the higher mechanical properties, reaching a hardness of 12.6 ± 1.2 GPa, compared to 5.3 ± 0.6 GPa of the NiFeCr matrix, and 4.3 ± 0.6 GPa of the tungsten phase. To diminish point-to-point variation of the measured data, at least ten nanoindentations were made in different points within the same phase, in accordance with the Oliver and Pharr method [4].

The load-displacement curves obtained from the different specific phases in both types of coatings are shown in Figs. 10.6 and 10.7, which also show two scanning probe microscopy images from two of the main phases. For the 310NS coating deposited by combustion flame spray process (Fig. 10.6), the load-displacement curves of the AlSi matrix are very homogeneous compared to those of the carbon phase, which have high scattering due to the low homogeneity of the material. The difference in mechanical properties between these phases is clearly visualized by the contrast of indentation mark sizes produced by the application of the same load. For the 140MXC coating deposited by arc spray process (Fig. 10.7), all load-displacement curves show low scattering and disparities among the three phases studied. The calculation of the elastic modulus for these phases allowed to determine that boron carbide particles yield the highest value E_r = 214.3 ± 13.2 GPa,

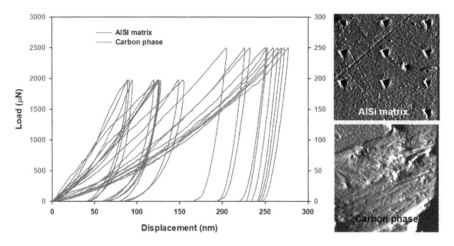

Fig. 10.6 Load-displacement curves and scanning probe microscopy images of the main phases from the 310NS coating deposited by the combustion flame spray process

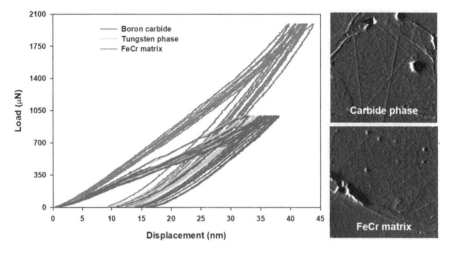

Fig. 10.7 Load-displacement curves and scanning probe microscopy images of the main phases from the 140MXC coating deposited by the arc spray process

compared to 126.1 ± 6.2 GPa from the NiFeCr matrix, and 109.7 ± 5.1 GPa from the tungsten phase.

Results of the tribological properties using the pin-on-disk test are given in Fig. 10.8a for the coefficient of friction and in Fig. 10.9 for the wear rate and volume loss of the thermal spray coatings and the 6061-T6 aluminum substrate. As can be seen in Fig. 10.8, the coefficient of friction is higher and more heterogeneous for the substrate compared with that of the coatings. Further, there are significant differences in the friction behavior with the distance of the arc spray coating compared to

10.2 Applications of Thermal Spray Coatings

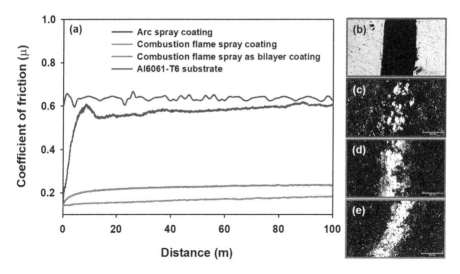

Fig. 10.8 (**a**) Coefficient of friction and wear track images of (**b**) Al6061-T6 substrate, (**c**) AS coating, (**d**) CFS coating, and (**e**) CFS as bilayer coating

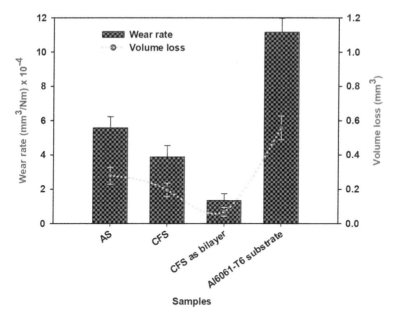

Fig. 10.9 Wear rate and volume loss of coatings deposited on the 6061-T6 aluminum substrate. AS, arc spray process; CFS, combustion flame spray process

Fig. 10.10 SEM images from the combustion flame spray coating worn surface using 5 N of load on a pin-on-disk test

the combustion flame spray coating. The latter shows the lowest and the most stable coefficient of friction, which is related to the effect of the coating microstructure and its chemical composition, mainly due to the presence of the carbon phase. Additionally, some surface photomicrographs showing the wear track after the samples were tested are shown in Figs. 10.8b–e. It is worth noting that at a normal load of 5 N, the wear rate of the Al6061-T6 is about eight and two times higher than the coating deposited by the flame combustion process and arc spray process, respectively. The incorporation of coatings on the aluminum alloy led in all the cases to reductions in both the wear rate and volume loss (Fig. 10.9). In the combustion flame spray (CFS) coating, the wear rate of the sample containing a high quantity of carbon phase was calculated as 3.9×10^{-4} mm^3/Nm. This phase acted as a lubricant together with the AlSi soft matrix. The AlSiC coating as bilayer, e.g., deposited on the FeCrBSiNbW coating, shows the minimum wear rate of the samples, with a value of 1.36×10^{-4} mm^3/Nm. The wear rate of the arc spray coating sample was higher, reaching a value of 5.6×10^{-4} mm^3/Nm. FeCrBSiNbW arc spray coating increases the wear resistance compared to the Al6061-T6 substrate, due to its plasticity and the confinement of cracks on the amorphous phase. The best wear resistance of the AlSiC combustion flame spray coating is due to a soft material showing a minimal coefficient of friction.

With the aim of observing in a more detail the wear track in two types of coatings, some SEM images were taken using different magnifications. Figure 10.10 depicts the results of the wear track size on the worn surface from the combustion flame spray coating. The worn surface shows few pits, plastic deformation in the carbon phase, and cracks in AlSi matrix, which are indicated by arrows. As a result of this wear behavior, it can be considered that the coated system using this load worked in ultra-mild conditions and did not suffer any appreciable wear, as well as no detachment of the coating during the test. However, the presence of voids and independent splats in the coating may have favored the initiation and propagation of cracks and rippled patterns (Fig. 10.10c). The plastification of the coating seems to be responsible for the wear values measured. Transference of material from the coating to the pin was also observed, indicating the presence of adhesive wear mechanisms. This tribological adhesive behavior in the AlSiC coating caused the presence of friction crack joints, as observed in Fig. 10.10d. The results show that, in combustion flame spray deposited coatings, a higher content of oxide strings and porosity were observed compared to the arc spray coatings; however, in this case the carbon phase helped as a lubricant showing a better and lower coefficient of friction.

In the same context, Fig. 10.11 shows SEM images from wear track and worn surface using 5 N on pin-on-disk test for the 140MXC coating deposited by the arc spray process. In this case, the worn surface shows few pits, plastic deformation, and some grooves indicated by an arrow in Fig. 10.11b. The nature grooves were formed by plowing out the material. Figures 10.11c-d reveal the presence of interparticle cracks, debris, delamination, and rippled patterns, indicating in this case an abrasive wear mechanism. Delamination was the main wear process observed by SEM, with removal of big zones of the material ranging between 20 and 40 μm (Fig. 10.11c). This mechanism was responsible for the wear rates observed, although some fine aligned grooves could be also observed on its surface, indicating that abrasion was also taking place in this material. On the other hand, some debris was formed in this coating (Fig. 10.11c). The small white particles in the debris must have been created due to the subsurface crack formation, propagation, and delamination in the specimen. Surfaces that are loaded repeatedly, even if they display some plastic deformation in the initial stages of their history, can generate patterns of subsurface protective stresses, which are sufficiently strong to enable the applied loads to be carried entirely elastically in the long term after many applications of the load cycle (shaken down). Despite the relief of stress levels that this generates, surfaces that have this feature can still wear subsequently by processes of surface fatigue or related phenomena. Subsurface cracks may be nucleated at microstructural defects or inclusions in the material, giving rise to the characteristic pitting fatigue and generating wear or debris particles, which have more or less an equiaxial morphology. Where the friction or traction forces are sufficient to deform the surface layer, material may be lost in the form of thin flakes or platelets. In such delamination wear, subsurface plastic shear is associated with the formation and

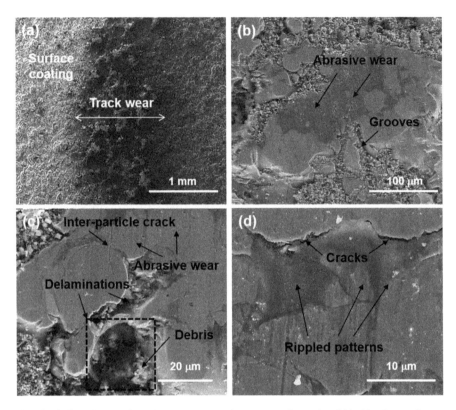

Fig. 10.11 SEM images from the arc spray coating worn surface using 5 N of load on a pin-on-disk test

propagation of cracks, nucleating from preexisting voids or inclusions present in the material structure.

10.3 Conclusions

With all the valuable research work conducted thus far in the field, there is no doubt that aluminum plays an important role in the development of enhanced materials used in industrial applications.

Regarding the case of thermal spray coatings, delamination in combination with plastic deformation and sparse oxidation was detected as wear mechanisms in the arc sprayed coating. The use of the arc spray process enables light alloys to be coated for applications requiring high hardness and low to moderate wear resistance. In terms of producing crystalline alloy coatings, it is, therefore, necessary to consider several factors including the presence of amorphous and/or crystalline structure in the original starting powder, as well as the spraying parameters employed which influence the crystallinity of amorphous phases in the coating due to reheating effects and cooling processes. The higher wear resistance of the coating compared to the uncoated substrate was achieved due to a combination of high hardness and thermal stability. On the other hand, the as-sprayed AlSiC coating revealed a high level of porosity, with low mechanical properties, despite presenting a better contact among the deposited splats and improved bonding at the substrate/coating interface. This type of AlSiC coating showed the best tribological properties, recording eight times less wear rate compared to the uncoated substrate. The wear mechanisms observed for this type of coating were adhesive behavior, plastification, and high presence of oxides in the coating. The carbon phase acts as a great lubricant, resulting in the lowest coefficient of friction and consequently the lowest wear rate. The features inherent in thermal spray processes and coating microstructures can be considered as both advantages and disadvantages, according to the application. Therefore, optimization of spray conditions and coating microstructures must be made by taking specific applications on components manufactured from light alloys and composites.

References

1. Varol Özkavak, H., et al. (2019). Comparison of wear properties of HVOF sprayed WC-Co and WC-CoCr coatings on Al alloys. *Materials Research Express, 6*(9), 096554.
2. Rokni, M. R., et al. (2015). An investigation into microstructure and mechanical properties of cold sprayed 7075 Al deposition. *Materials Science and Engineering A, 625*, 19–27.
3. Dayani, S. B., et al. (2018). The impact of AA7075 cold spray coating on the fatigue life of AZ31B cast alloy. *Surface and Coatings Technology, 337*, 150–158.
4. Oliver, W. C., & Pharr, G. M. (1992). An improved technique for determining hardness and elastic modulus using load and displacement sensing indentation experiments. *Journal of Materials Research, 7*(6), 1564–1586.
5. ASTM G99-17. (2017). *Standard test method for wear testing with a pin-on-disk apparatus.* West Conshohocken: ASTM.
6. Axinte, E. (2012). Metallic glasses from "alchemy" to pure science: Present and future of design, processing and applications of glassy metals. *Materials & Design, 35*, 518–556.

Index

A
Additive manufacturing (AM), 7
    capabilities, 90
    conventional manufacturing techniques, 89
    cost reduction, 92
    environment, 93
    hybrid systems, 93
    laser systems, 92
    metal casting, 90
    metallic materials, 89
    multiple layers of material, 89
    multiple materials, 93
    prototype production, 90
    prototyping process, 89
    remanufacture, 94
    zero waste, 93
Aeronautic industry, 1
AlSiCuCeFeMnNi alloy, 27
Aluminum (Al_2O_3), 5, 13, 18
Aluminum alloys, 1, 2, 51, 112
Aluminum carbide (Al_4C_3), 168
Aluminum composites, 173
    MWCNT content, 171
Aluminum-lithium (Al-Li) alloys, 3
Annular dark-field (ADF) technique
    bright and dark field images, 175
Arc spray (AS), 108, 109, 187
Archimedes method, 129, 132, 133, 162
Atomic absorption spectrometry (AAS), 130
Atomic force microscopy (AFM), 159–161, 163
AZ31B-MWCNTs composites, 153, 154

B
Backscattered electron (BSE) signal, 79
Beryllium alloys, 2, 4
Boron (B), 5
Boron-and carbon-epoxy composites, 1
Bulk mechanical properties, 65
Bulk-scale mechanical testing
    structural applications, 152
    tensile-compression tests, 152
    tribological evaluation, 152, 153

C
CALculation of PHAse Diagram (CALPHAD)
    methodology approach, 22
Cantilever beam test, 161, 162
Carbon nanotubes (CNTs), 5, 40–43, 45, 49, 168, 169
Characterization techniques, 8, 9
    chemical analysis, 130
    density measurement, 131–136
    EDS, 140, 142
    EELS, 149
    elemental chemical analysis, 129
    HRTEM, 147, 149
    light alloys and composites analysis, 129
    materials composition, 129
    OM, 136
    Raman spectroscopy, 138–140
    SEM, 140
    structural analysis, 130
    TEM, 143–146
    thermal analysis, 131
    types, 130, 162
    XPS, 150
    XRD, 136, 138, 162
Chemical analysis, 130
Chemical composition, 115

Chemical vapor deposition (CVD), 5, 41
CHNS-O elemental analyzer, 130
Coating wear rate, 186
Coefficient of friction, 191
Cold spray, 110–112
Cold spray low pressure (CSLP)
   industrial sector, 98
   light alloys, 98
   metal components, 98
   oxide-free deposits, 98
   powders, 99
   process and equipment, 98, 99
Combustion flame spray (CFS), 107, 108, 112, 119, 120, 187
Compaction process, 34, 37, 38
Computational fluid dynamics (CFD), 97
Computed tomography (CT) scanning, 135
Concentric cylinder model, 140
Conventional light metallic alloys, 2
Critical resolved shear stresses (CRSSs), 178
Crystalline alloy coatings, 194

D
Density, 131
Density measurement
   Archimedes' method, 132, 133
   CT scanning, 135
   pychometry, 133, 134
   weight-to-strength ratio, 131
Differential scanning calorimetry (DSC), 131
Differential thermal analysis (DTA), 131
Diffusion bonding mechanism, 54, 55, 57
Direct metal laser sintering (DMLS), 90
   complex 3D architectures, 94
   plastic injection mould industry, 97
   powders, 95, 96
   process and equipment, 95, 96
Dispersion quantification, 61, 62

E
Electron energy loss spectroscopy (EELS), 149, 179, 180
Electropolishing, 143
Energy-dispersive spectroscopy (EDS), 17, 28, 59, 72, 81, 84, 115, 140, 142, 171, 179
Equilibrium phase diagram, 71

F
Fast Fourier transform (FFT), 175–176
Field emission scanning electron microscopy (FE-SEM), 58, 114–115
Flame atomic absorption spectrometry (FAAS), 130
Flame spray coating, 192
Focused ion beam (FIB), 63, 84, 144, 145, 161, 171
Fracture toughness, 162
Full width at half maximum (FWHM), 137
Functionalized carbon nanotubes (f-CNT), 159
Functionalized graphene (f-G), 159

G
Graphene (GNp), 159

H
Hcp unit cell, 177
He-Ne laser light, 138
High-frequency plasma or induction systems (HFIS), 40
High-pressure torsion (HPT), 69, 70, 131
High-resolution transmission electron microscopy (HRTEM), 147, 149, 163, 174, 176–178, 180
High Score Software®, 115
High velocity oxy-fuel (HVOF) process, 109
Hybrid systems, 93

I
In situ characterization techniques, 163
Inductively coupled plasma (ICP)
   analysis, 8
Inductively coupled plasma mass spectrometry (ICP-MS), 130
Inductively coupled plasma optical emission spectrometry (ICP-OES), 130, 162
Industrial applications
   aerospace structures, 183
   Al-2024, 184
   Al-6082, 184
   Al-7075, 184
   aluminum alloys, 183
   composite properties, 184
   light alloys, 184
   magnesium (Mg) alloys, 183
Interface, 167–170
   Sandwich technique, 171
Interface strengthening mechanisms, 169
International Centre for Diffraction Data (ICDD), 115
Interplanar distance, 177

Index

L
Light alloys, 5, 9, 10
  arc spray process, 108, 109
  CFS, 107, 108
  cold spray, 107, 110–112
  HVOF process, 109
  industrial applications, 106
  microstructure, 107
  spray materials, 112–115
  warm spray, 110–112
Light alloys and composites
  mechanical characterization, 163
  mechanical properties
    Bulk-scale mechanical testing, 152–155
    experimental issues, 151
    testing (*see* Nano- and micromechanical testing)
Longitudinal face, 170

M
Magnesium, 18
Magnesium alloys, 3, 50, 103, 112, 168
Magnesium composites, 173
Magnesium oxide (MgO), 168, 172
Manufacturing process
  accelerated alloy design, 21
  aerospace industries, 18
  Al7075 alloy, 24
  Al-Mg phase diagram, 23
  Al-Si matrix-based light alloy, 27
  Al-Si phase diagram, 26
  aluminum, 13, 14, 18
  CALPHAD, 22
  casting process, 14
  chilling system, 14
  commercial and open-source software, 23
  computational thermodynamics, 23
  cooling curves, 16
  Gibbs energy, 22
  grain refinement mechanism, 15
  heat treatments, 13
  light alloys, 14
  lightweight materials, 13
  magnesium, 14, 18
  metallic alloys, 21
  Mg-Al phase diagram, 18, 19
  microstructures, 14, 16, 17
  Ni nanoparticles, 17
  SEM, 29
  TEM, 25, 26
  Ti phase, 17
  XRD patterns, 25

Matrix-reinforcement interface
  surface roughness, 170
Mechanical alloying (MA), 6, 33, 40–43, 45
Mechanical characterization, 151, 152, 162, 163
Mechanical milling (MM), 6, 34, 36, 41, 43–46
Mechanical properties, 9, 49, 50, 52, 58, 61, 64, 65
Metal matrix composites (MMCs), 2, 4, 19, 49, 112, 167, 171
Metal matrix composites synthesis
  aluminum alloy, 54
  AZ31B magnesium alloy, 53
  bulk mechanical properties, 64, 65
  diffusion bonding mechanism, 54, 55, 57
  dispersion quantification, 61, 62
  microstructural and structural analysis, composites, 58–60
  microstructural changes, 53
  microstructural evolution, 62, 64
  optical microscopy, 55
  polymer matrix composites synthesis, 51, 52
  reinforced zones, 54
  temperature, 54
  thermogravimetric analysis, 54
Metallic matrix, 168
Microstructural behavior, 163
Microstructure
  Al-Si matrix, 119
  AlSiC coating, 119
  amorphous and crystalline phases, 120, 123
  amorphous-crystalline composite coating, 120
  arc spray process, 116, 121
  BSE-SEM micrographs, 117
  carbon, 117
  chemical composition, 120
  coating process, 118
  EDS elemental analysis, 123
  flame thermal spray process, 116
  gas pressure ratio, 116, 117
  intra-splats cracks, 119
  lamellar structures, 117
  measurement spots, 119
  morphology, 122
  non-smooth surface, 115
  oxygen, 123
  phases, 123
  SEM images, 116, 118, 122
  SEM micrographs, 115
  SEM-EDS analysis, 122

Microstructure (*cont.*)
 spray dry technology, 115
 surface morphology, 119
 XRD patterns, 117, 120
Modulus mapping, 156–158
Multi-walled carbon nanotubes (MWCNTs), 171, 173–180

N
Nano-and micromechanical testing
 AFM, 159
 cantilever beam test, 161, 162
 modulus mapping, 156, 157
 nanoindentation techniques, 155
 nanoscratch, 157
Nanocomposites
 aluminum-matrix composites, 19
 SLM, 20
 SSM, 20
 stir casting, 20
Nanocrystalline phases, 108
Nanoindentation, 155, 156, 159, 185
Nanoscratch test, 157, 158

O
Optical microscopy (OM), 72, 136, 137

P
Phase formation (interphases), 168
Phase transformation, 77, 78, 83, 85
Pin-on-disk test, 192
Plasma electrolytic oxide (PEO), 136
Plastic injection mould industry, 97
Polymer matrix composites synthesis, 51, 52
Polyvinyl alcohol (PVA), 52
Powder Cell software, 74
Powder metallurgy (PM), 6
 advantages, 35
 application, 33
 atomization, 33
 CNTs, 40–43, 45
 cold compaction process, 37, 38
 degree of compressibility, 37
 functional material, 33
 light alloys, 35, 36
 liquid routes, 33
 lubricants, 37
 materials manufactured, 33
 mechanical milling, 34
 production process, 34
 SE-SEM micrographs, 34
 single compaction process, 37

sintering process, 34, 38–40
unconventional techniques, 33
Pressure-temperature (P-T) phase diagram, 74
Process control agent (PCA), 77
Pycnometry, 133, 134

R
Raman spectroscopy, 9, 138–140
Recrystallization, 173
Reinforcement, 167, 169, 170
Reinforcing materials, 4
Residual stresses, 169
Resistance-weight ratio, 1
Rietveld method, 137

S
SAED pattern, 145, 146
Sandwich technique, 6, 7
 aluminum alloys, 51
 CNTs, 49
 compaction and rolling processes, 49
 layered composites, 49
 light alloys, 50
 magnesium alloys, 50
 MMCs, 49
 MWCNTs, 50
 processing techniques, 49
Scanning electron microscope (SEM), 28, 72, 79, 140, 144, 163, 171
Scanning force microscopy (SFM), 159
Scherrer formula, 136
Screened X-ray energy, 140
Secondary electron (SE) signal, 79
Secondary electron scanning electron microscopy (SE-SEM) images, 73
Selected area electron diffraction (SAED) patterns, 83–84
Selective laser melting (SLM), 20, 90
SEM-EDS elemental mapping analysis, 142
Semisolid metal (SSM) method, 20
Severe plastic deformation (SPD), 7
 characterization, Ti-Mg alloys, 76–79, 81–85
 HPT, 69, 70
 metal forming, 69
 Ti-Mg alloys, 72, 74–76
 titanium-magnesium alloys, 71, 72
Silicon carbide (SiC), 5
Single-walled carbon nanotubes (SWCNTs), 169
Sintering process, 34, 35, 37–40
Solid-state metal powders, 41
Spraying process parameters, 115

Index

Stir casting, 20
Stress transfer
　CNT ends, 169
　CNT surface, 169
　wettability, 167
Stress transmission, 169

T
Taylor factor, 170
TEM sample preparation
　electropolishing, 143
　image acquisition, 143
　ultramicrotomy, 144, 145
Tensile-compression tests, 152
TGA-DSC analysis, 131, 132
Thermal analysis, 131
Thermal spray coatings, 8
　advantages, 104
　AlSi matrix, 189
　AlSiC, 185
　aluminum alloys, 103
　amorphous alloy, 188
　application, 104
　chemical reactions, 104
　deposition technique, 106
　engineering parts, 104
　FeCrBSiNbW, 185
　flame spray process, 190
　geometry, 104
　heater/power source, 104
　high-velocity droplets, 104
　industrial sectors, 186
　light alloys, 103
　lower costs, 185
　magnesium alloys, 103
　microhardness, 187
　140MXC coating, 189
　nanocrystalline alloy, 188
　nanoindentation, 188
　NiFeCr matrix, 190
　310NS coating, 187
　operating principle, 104
　particles, 104
　physical and chemical phenomena, 105, 106
　post-spray sealing, 105
　powder materials, 105
　process parameters, 104
　quality coatings, 185
　residual stresses, 106
　spray process, 190
　　Vickers microhardness (HV), 188
　　wear rate and volume loss, 190, 191
Thermal spray processes, 113–114
Thermogravimetric analysis (TGA), 131
Ti-Mg alloys, 72, 74–76
Titanium alloys, 3
Titanium-magnesium alloys, 71, 72
Transmission electron microscopy (TEM), 24, 63, 72, 83, 145, 146, 149, 161, 163, 173–175, 180
Tribological evaluation, 152, 153

U
Ultramicrotomy, 144

V
VESTA software, 177
Vickers microhardness, 159, 160, 185

W
Warm spray, 110–112
Wear mechanisms, 194
Wear track images, 191
Wettability, 169

X
X-ray diffraction (XRD), 24, 72, 76, 136, 138, 162, 174
X-ray/neutron imaging, 132
X-ray photoelectron spectroscopy (XPS), 9, 150, 163

Y
Young's modulus, 140

CPSIA information can be obtained
at www.ICGtesting.com
Printed in the USA
LVHW081751290621
691473LV00001B/52